The Organic Waste Composting Handbook

The Organic Waste Composting Handbook

Editor: Jackson Lawrence

MURPHY & MOORE
www.murphy-moorepublishing.com

Murphy & Moore Publishing,
1 Rockefeller Plaza,
New York City, NY 10020, USA

Visit us on the World Wide Web at:
www.murphy-moorepublishing.com

ISBN: 978-1-63987-530-6 (Hardback)

Cataloging-in-Publication Data

The organic waste composting handbook / edited by Jackson Lawrence.
    p. cm.
Includes bibliographical references and index.
ISBN 978-1-63987-530-6
1. Organic wastes. 2. Compost. 3. Refuse and refuse disposal.
4. Salvage (Waste, etc.). I. Lawrence, Jackson.
TD796.5 .O74 2022
631.875--dc23

Table of Contents

Preface...VII

Chapter 1 **Biochar–Compost Mixtures as a Promising Solution
to Organic Waste Management within a Circular
Holistic Approach**...1
Federico Varalta and Jaana Sorvari

Chapter 2 **Composting: A Sustainable Route for Processing
of Biodegradable Waste in India** 21
Ashootosh Mandpe, Sweta Kumari and Sunil Kumar

Chapter 3 **Organic Waste Composting at Versalles: An
Alternative that Contributes to the Economic,
Social and Environmental Well-Being of
Stakeholders**..43
Luis Fernando Marmolejo-Rebellón,
Edgar Ricardo Oviedo-Ocaña and
Patricia Torres-Lozada

Chapter 4 **Organic Waste Composting through Nexus
Thinking: Linking Soil and Waste as a Substantial
Contribution to Sustainable Development**............................60
Hiroshan Hettiarachchi, Johan Bouma,
Serena Caucci and Lulu Zhang

Chapter 5 **Composting as a Municipal Solid Waste
Management Strategy: Lessons Learned from
Cajicá, Colombia**..75
Cristian Rivera Machado and Hiroshan Hettiarachchi

Chapter 6 **Composting in Sri Lanka: Policies, Practices,
Challenges and Emerging Concerns**.......................................97
Warshi S. Dandeniya and Serena Caucci

Chapter 7 **Co-composting: An Opportunity to Produce
Compost with Designated Tailor-Made Properties** 126
Laura Giagnoni, Tania Martellini, Roberto Scodellini,
Alessandra Cincinelli and Giancarlo Renella

Chapter 8   **Urban Waste as a Resource: The Case of the**
                Utilisation of Organic Waste to Improve
                Agriculture Productivity Project..153
                Dzidzo Yirenya-Tawiah, Ted Annang,
                Benjamin Dankyira Ofori,
                Benedicta Yayra Fosu-Mensah,
                Elaine Tweneboah-Lawson, Richard Yeboah,
                Kwaku Owusu-Afriyie, Benjamin Abudey,
                Ted Annan, Cecilia Datsa and Christopher Gordon

Chapter 9   **Valuing Waste – A Multi-method Analysis of the**
                use of Household Refuse from Cooking and
                Sanitation for Soil Fertility Management in
                Tanzanian Smallholdings...176
                Ariane Krause

Chapter 10 **Traditional and Adapted Composting Practices**
                Applied in Smallholder Banana-Coffee-Based
                Farming Systems: Case Studies from Kagera and
                Morogoro Regions, Tanzania...208
                Anika Reetsch, Didas Kimaro, Karl-Heinz Feger and
                Kai Schwärzel

Permissions

List of Contributors

Index

Preface

Any waste that can be broken down into simple organic molecules like water, methane and carbon dioxide by microorganisms is termed as organic waste. Breakdown of organic waste involves processes such as aerobic digestion, anaerobic digestion and composting. Human waste, sludge, manure, food waste, etc. are some common examples of organic waste. Composting is the process of decomposing organic waste into humus rich soil known as compost. It is an aerobic process which requires an adequate supply of carbon, nitrogen, oxygen and water for microorganisms to work efficiently. From theories to research to practical applications, case studies related to all contemporary topics of relevance to this field have been included in this book. Most of the topics introduced herein cover new techniques and the applications of organic waste composting. This book is an essential guide for both academicians and those who wish to pursue this discipline further.

This book is a result of research of several months to collate the most relevant data in the field.

When I was approached with the idea of this book and the proposal to edit it, I was overwhelmed. It gave me an opportunity to reach out to all those who share a common interest with me in this field. I had 3 main parameters for editing this text:

1. Accuracy – The data and information provided in this book should be up-to-date and valuable to the readers.

2. Structure – The data must be presented in a structured format for easy understanding and better grasping of the readers.

3. Universal Approach – This book not only targets students but also experts and innovators in the field, thus my aim was to present topics which are of use to all.

Thus, it took me a couple of months to finish the editing of this book.

I would like to make a special mention of my publisher who considered me worthy of this opportunity and also supported me throughout the editing process. I would also like to thank the editing team at the back-end who extended their help whenever required.

Editor

Biochar–Compost Mixtures as a Promising Solution to Organic Waste Management within a Circular Holistic Approach

Federico Varalta and Jaana Sorvari

Abstract In the common linear economy approach, organic waste treatment mainly generates energy, due to the existing demand and the goal of reducing the use of fossil fuel. Yet recent innovations and associated products are calling for an increasingly diverse use of organic waste within a circular holistic framework where the biochar and composting mixture appears to be the key to achieving a robust solution for sustainable development. Nonetheless, the inhomogeneity of organic waste and the synergies between biochar and composting require further investigation before broad-scale field application. In this chapter, we illustrate how governmental policies should be updated and revised to effectively support the development of new sustainable solutions, that should take into account social, economic and environmental implications, as well as their mutual interactions. As a consequence, robust tools and reliable procedures to evaluate sustainability will have to be established in this new ecological structure.

Keywords Organic waste · Waste management · Biochar–compost mixtures · Terra preta · Nutrients recycling · Circular economy · Sustainability

1 Introduction

Global population growth and the increased consumption of natural resources are placing an unprecedented pressure on already strained natural resources. To cope with increased food demand, human activities related to agriculture and food production have significantly altered nutrient cycles, depleted soils and increased greenhouse gas (GHG) emissions. Global population growth and the related food

F. Varalta (✉) · J. Sorvari
Aalto University, Espoo, Finland
e-mail: federico.varalta@aalto.fi; jaana.sorvari@aalto.fi

demand are straining the already fragile equilibrium of natural resources. The new sector of bioeconomy is seeking to reconcile the challenges of producing more food, while, at the same time, lowering the environmental burden. In this context, the proper management of organic waste can play a prominent role in tackling the mounting challenges. At the same time, current recycling and rebalancing actions have been insufficient to counter these adverse environmental consequences. Nutrients and organic matter have leaked without any control into the environment through food processing activities, excessive crop fertilising, poorly managed breeding of livestock and uncollected human waste (Kirschenmann 2010). The results of such anthropogenic pressure on natural resources have turned the ecosystem into a suffering status: with eutrophication of water bodies, air pollution and global warming. In addition, despite the higher food production capacity achieved, one person out of seven still has limited access to food and suffers from hunger. This, in turn, has exacerbated social discrimination and increased inequality (Foley et al. 2011). The concepts of bioeconomy (first) and circular economy (later) were developed to tackle these challenges, and to reduce the burden on natural resources. Many sectors of the economy are reassessing the way they operate, with the aim of achieving a sustainable mode of action. As a consequence, material flows in past years have been simplified and new methods have been developed for efficient reuse, or optimisation of the use, of natural resources. Fast-changing sectors of the global economy, such as energy production, have implemented circular economy, while more traditional segments, for example food production and agriculture, are still lagging (Buckwell and Nadeu 2016). Likewise, actions have been taken and the model of circular economy promoted across all sectors, fuelled also by updates in international and local legislation. To date, several governments and international organisations have developed programmes to improve the sustainability of human activities (European Commission 2012, 2018; The White House 2012), and governmental institutions have started to prompt policies to facilitate the shift towards circular economy. Educational organisations (Ellen Macarthur Foundation 2019; UN Food and Agriculture Organization (FAO) and RUAF Foundation 2015) are promoting research to overcome the deficit of circular economy in traditional sectors, and gradually new notions of bioeconomy are being shaped. Ultimately, bioeconomy offers the possibility to develop new sustainable processes that make it possible to secure the world's future food production demand while significantly reducing the environmental burden of related activities. Under these circumstances, the sustainable management of organic waste through efficient recycling of nutrients contained in that waste, increased carbon sequestration, and green energy production plays a prominent role.

In this chapter, the concept of a circular holistic approach to biowaste management is introduced. A critical comparison between the linear and the circular approach is presented, with a particular focus on the potential benefits of biochar–compost mixture (BCM) in the global context of bioeconomy.

In the long run, the aim is not only to preserve natural resources and to create an environmentally and economically sustainable future, but also to create a viable economy sector that will gradually replace obsolete courses of action and create new job opportunities. In other words, a complete shift of mindset is required in the implementation of bioeconomy, where traditional methods and treatment techniques are deemed no longer applicable and new ones must be developed. The circular holistic approach to organic waste management offers the opportunity to answer the new challenges. In particular, the development of techniques that efficiently return nutrients and organic matter back to the soil can be one of the keys to a new sustainable future.

2   Organic Waste in a Linear Economy

Despite the recent effort in terms of policies to promote the reuse and recycling of organic waste in different forms, the biowaste management sector still largely relies on a straightforward approach, with a linear progress.

The linear approach consists mainly of applying the same treatment process to all available organic waste streams, regardless of their characteristics. Such an approach usually results in a main product with high profitability and a side product that, most of the time, has no—or negligible—commercial value. In commercial terms, the offered solution has the sole benefit of maximising the output quantity of a single product. Currently, this approach is preferred over more complex schemes because, in most cases, the end products already have an established market, with existing integrated logistics and an appropriate distribution network. In fact, these end products are typically drop-in commodities (Webster and Francis 2016), such as fuels or energy, with no need to develop new markets for alternative products. These linear processes are also robust and well established among the organic waste treatment industry, have become more reliable, do not pose any particular challenges in technical or operational terms, and are well framed in the environmental legislation.

Nevertheless, it is undeniable that the simplicity of the linear approach also represents its major drawback. As indicated by the 17 Sustainable Development Goals issued by the United Nations (United Nations 2018), sustainability is a complex subject that cannot be simplified to the implementation of one single, though important, target, in tackling global sustainability challenges. Aside from the social

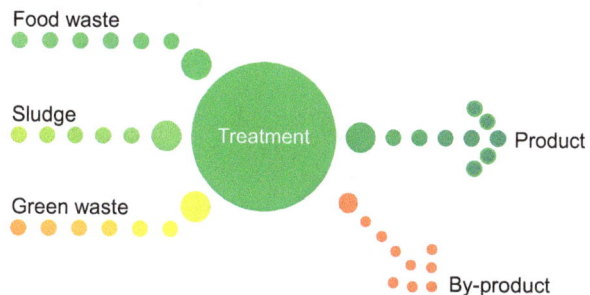

Fig. A typical linear process for organic waste treatment

Food waste

Sludge

Treatment

Green waste

Product

By-product

aspects of the sustainability goals, the integration of other environmental criteria besides carbon and water footprints are of paramount importance in order to secure the sustainability of recycling processes as well as the related value chains. More specifically, organic waste treatment processes should be evaluated also from the viewpoint of rebalancing the nutrients flows in ecosystems and the potential to sequestrate carbon in the soil.

3   Novel Mode of Action Founded on Circularity

The development of circular models and the associated public awareness of sustainability makes the linear approach obsolete. It is therefore necessary to elaborate new models that allow the full use of organic waste potential, and, at the same time, secure overall sustainability. A circular approach that takes into account the multifaced aspects of organic waste treatment, that is a holistic approach, offers the possibility to answer the mounting challenge by integrating new treatment technologies with more traditional processes.

The holistic circular approach shares some traits with the industrial symbiosis concept, where a network of separate industries recovers and redirects resources and the by-products of its processes to improve the efficiency of the overall business. In the case of organic waste management, the holistic circular approach would suggest that we first treat the biological fraction of municipal solid waste (MSW) to produce, for example, animal feed through the cultivation of larvae insects. The remaining unused substrate could be combined with liquid organic waste from dairies in a wet biogas process and the digestate could then be combined with biochar produced from the pyrolysis of scrap wood or any other waste with a high content of lignin and composted to produce a soil amendment or an organic fertiliser. This type of approach is extremely flexible as it can be adapted to virtually any

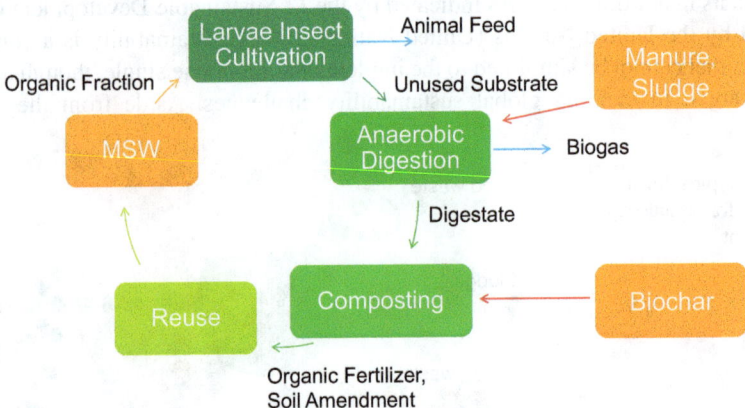

Fig. Framework of a holistic circular approach to the management of organic waste

conditions. In fact, there are several combinations of the processes for treating organic waste and the development of new applications increases the number of possibilities. The holistic methodology behind the circular approach allows for the introduction of the cascade organic waste treatment concept. This model originated in the paper industry, where higher value-added products are first extracted, and at the end of the product transformation cycle residues of the separate treatments are redirected to energy production processes (Keegan et al. 2013). For organic waste, the components with higher commercial value would be initially separated, while the residues or by-products from the process that generates the high-end products would form the basis for producing organic fertiliser. Through this theoretical model, the natural cycle of nutrients and organic matter might be completely closed.

The approach of circularity in organic waste management offers a number of advantages. Primarily, the heterogenous organic waste streams can be rationalised by directing them to the most suitable treatment process, reducing the risk of losing efficiency. The characterisation of organic waste is also an important factor for process efficiency implementation, as each treatment process is selected in accordance with the raw material characteristics. It thus becomes clear that this process chain can better tolerate higher variations in the organic waste quality and, consequently, larger fluctuations in the end products' market price. Such an approach no longer depends on the availability of a single feedstock or a few feedstocks, and it can be easily adjusted according to the local conditions. In addition, it enables quick adaptation in the event of changes in the demand of the products at the global level. A consequence of the higher process flexibility of the circular approach is the possibility of bridging the gap between the multiple sectors generating organic waste and creating a network where mutual added value is generated. In the past, the lack of integration of waste management process, the use of specialised technologies and the legislative issues of sectors like agriculture, sanitation, or MSW have led to a very fragmented and inefficient environment for by-product production (Buckwell and Nadeu 2016). The intrinsic flexibility of the holistic approach carries within itself the opportunity to further streamline the production processes of single operation by creating an additional recycling loop between the different sectors and consequently generating more added value.

In addition to the fact that the integration of different sectors contributes to improve efficiency of organic waste treatment, the recycling of nutrients is optimised, and nutrient losses are reduced (Sutton et al. 2013). Nevertheless, despite the recent improvement in managing organic waste streams, nutrients still leak into the environment.

Contrary to the linear approach, the circular approach makes it possible to focus on nutrients reuse and recycling across the whole range of organic waste. The improved management of nutrients entails the identification of the proper actions for a specific nutrient and the selection of the most appropriate process in accordance with the final application.

The holistic circular approach offers the opportunity to improve carbon sequestration in soil. In particular, composting combined with biochar from the pyrolysis process enables the prospect of storing large quantities of carbon in soil as these

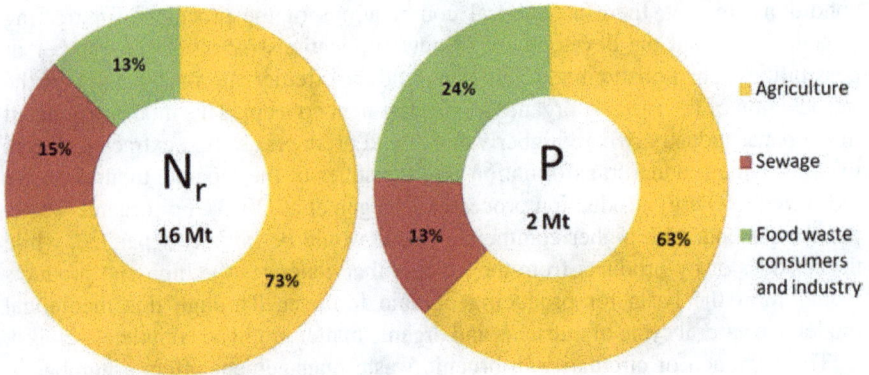

Fig. Main nutrients leakage from EU countries. (Buckwell and Nadeu 2016)

techniques can treat problematic biomasses and by-products from other treatment processes that would otherwise remain unutilised (Fischer and Glaser 2012; Viaene et al. 2016). Notably, the proper management of soil organic matter in terms of carbon sequestration contributes directly to the better management of the nutrients as it is the largest reservoir of nutrients. Hence, practices that improve carbon sequestration in soil will probably also have a positive impact on the availability of nutrients (Fischer and Glaser 2012).

3.1 Composting and Biochar: Synergism as a Support to the Holistic Approach

In a circular holistic approach context, the final step plays a very impor-tant role. The effective management of the final residuals from different processes guarantees not only the profitability of the process in economic terms, but also enables sustainability for human-related activities—in particular, agriculture and food production. In fact, the current shift in agriculture practices towards selective farming and knowledge-intensive precision crops requires an efficient n utrients recycling system, coupled with environmentally friendly practices, in order to return organic matter to the soil.

In this context, the qualities of the Terra preta (dark earth) agro-ecosystem offers a model to optimise organic waste management within a circular holistic approach. In fact, Terra preta is a particular soil of anthropogenic origin found in irregular patches in the tropical rainforests of Amazonia, characterised by high concentration of nutrients and high stability of humified organic matter. Terra preta was produced by the incorporation of charred human waste, which is rich in nutrients, in soil (Glaser and Birk 2012), and different studies have demonstrated the key role of biochar in the formation of dark earths over the years.

Biochar is obtained through pyrolysis by using organic waste as feedstock. Virtually any type of organic waste can be used as a raw material to produce biochar, although the type of feedstock significantly affects the properties of the produced biochar. According to the so-called 'charcoal vision' (Laird 2008), biomass processing through pyrolysis can be a sustainable approach for extracting energy from organic waste through their thermal transformation into bio-oil, syngas and charcoal. While the produced syngas can provide energy to the pyrolyser, the oil and charcoal fractions can be used too as secondary raw materials in various sectors. For example, the use of biochar as soils amendment can increase its physical, chemical and biological fertility, and represents a potential carbon sink because the estimated half-life of carbon in soil is more than 1000 years. Recently, biochar has gained renewed attention as a means to sequestrate carbon and, at the same time, improve soil quality (Tammeorg et al. 2014). It has already been observed that the direct application of biochar to the soil increases water retention capacity, total porosity, and soil mean temperature, and also improves the rate of ion exchange and retention of nutrients. An increased microbial activity has also been detected after the application of biochar in the soil (Gul et al. 2015), although the results from about a decade of experiments are not always consistent.

Even though biochar improves soil quality and the availability of nutrients, it does not provide soil with a sufficient amount of nutrients. The natural approach of combining biochar with compost instead produces a substrate rich in nutrients. The aim is, in the first place, to replicate the Terra preta process with the help of modern composting technologies and, successively, to produce an organic fertiliser or soil amendment as efficient as the current commercially available soil amendments and completely replace chemical fertilisers. In accordance with the circular holistic approach, the final target for biochar–compost use is thus the efficient management of available organic waste streams and a full exploitation of waste potential that generates sustainable added value in the existing production chains.

Nonetheless, the process of artificially recreating the Terra preta effect cannot be obtained by the simple combination of charred material and compost: It requires a deep understanding of the physico-chemical processes involved. The simple combination of charred material and compost does not result in the realisation of Terra preta. There may be many mechanisms that play a key role that have not yet been fully identified. In addition, the inhomogeneity of organic waste used is a challenge. The high variability in the quality of the raw material is not ideal for achieving a stable and reliable end product with a specific quality. Moreover, it should also be remembered that the aim is to employ the artificially produced Terra preta on soils that have different characteristics for various crops, and it cannot be excluded that artificial Terra preta with given characteristics may even produce contrasting results when applied on different soils.

In the next section, the initial promising findings of the interaction between biochar and composting are presented. The results from the land application of newly produced Terra preta are then separately evaluated.

3.2   Recent Developments in BCM

Integration of biochar in compost to simulate the realisation of artificial Terra preta soils is still at the early stage, and therefore the results of recent studies are indicative, not definitive, regarding the potential that lies in the synergies between biochar and raw material processed by composting. It is thus premature to assess if BCM fulfils the 'end of waste' criteria introduced in the Waste Framework Directive (2008/98/EC). In fact, no common methodology exists for the application of biochar to compost. So far, it has been applied indistinctively both before and after the composting process. Here, we introduce the results from studies where biochar has been applied before or during the composting process, because these procedures increase the benefits of the combination. Moreover, these specific set-ups allow us to evaluate how the composting parameter modification impacts the quality and the characteristics of the mixture of compost and biochar.

One of the most important additional benefits of adding biochar before the composting process is the temperature increase during composting. This is an aspect of great importance within the circular holistic approach. In fact, higher temperatures also mean higher levels of hygienisation, and allow the use of feedstocks, such as municipal sewage sludge and animal manure. It seems that the increase of the temperature is caused by the fact that biochar is filling the gaps in between the composting mass, thereby reducing the overall heat loss of the system (Zhang et al. 2014). This has a direct impact on the thermophilic phase of the composting process. It is interesting to note, however, that in some cases the addition of biochar extends the duration of the thermophilic phase (Li et al. 2015; López-Cano et al. 2016), while in other cases, it reduces the duration itself (Awasthi et al. 2016b). Some contradictory outcomes are most probably due to the temperature of the pyrolysis and the type of feedstock employed in biochar production (Awasthi et al. 2016a).

The higher retention capacity of the biochar also has a positive effect on the composting process. Moisture is a key parameter of the composting process, in order to obtain a mature compost with stable organic matter. Researches have registered higher moisture contents in composting piles with biochar when compared piles without biochar (Li et al. 2015). The higher moisture content in the pile consequently reduces the water losses and the leakage of nutrients (Theeba et al. 2012). Hence, biochar addition to the composting piles is indirectly responsible for more effective nutrient retention.

One of the main targets of the integration of the biochar and composting process is the increase of the content of nutrients and their bioavailability in the final product. While there is evidence of increased content of nutrients, the mechanisms involved are very complex and still not known in depth. Some studies have observed higher concentrations of P, K, Ca and Mg ions in the end product when biochar is added prior to the composting process (Zhang et al. 2014), when compared to end products without biochar, and this can be ascribed to the biochar's high ion exchange capacity. The negatively charged surface of the biochar has the ability to retain cations like K^+, Ca^{2+} and Mg^{2+}, through electrostatic attractions reducing the leaking during the composting process. On the other hand, the fate of phosphorus measured as phosphate (PO_4^{3-}) after biochar addition is not clear. Zhang et al. (2014) have ended up with contradictory results, showing that addition of biochar increases the PO_4^{3-} concentration in the end product when compared to the compost without addition, whereas during the process the total PO_4^{3-} concentration decreased. This was explained by hypothesising that soluble P could be taken up by the microorganisms during the composting process. However, the influence of biochar addition to the P bioavailability seems to be a more complex issue than has previously been observed (Vandecasteele et al. 2016; Xu et al. 2013). The initial P content in the biochar itself appears to be an important factor in determining the bioavailability of phosphorus in soil. In addition, the pH value seems to be a key element in determining the concentration of P in compost, but its measurement is complicated due to the interactions with other mineral compounds. Other studies also suggest that timing of biochar addition to the composting process affects the final concentration (Vandecasteele et al. 2017). The lack of consistent data on P concentration in the compost after biochar addition deserves further research in order to better understand the mechanism involved.

The addition of biochar to the composting process has clear beneficial effects on other macronutrients than P. Earlier studies have shown that the addition of biochar reduces N losses, owing to the capacity of biochar to adsorb ammonia (NH_4^+) and ammonium (NH_3) (Hua et al. 2008). However, this is not the only mechanism triggered by biochar in connection to N_2 emission reduction during the composting process. Lopez-Cano et al. (2016) reported that the addition of biochar to the composting process decreased the rate of the ammonification process. This result was explained by the fact that biochar creates a favourable environment for the growth of nitrifying bacteria, leading to the reduction of gaseous NOx emissions from the composting process, resulting in the double benefit of increasing the end product's value and reducing the GHG emissions in the atmosphere. This hypothesis is supported by the high porosity of the biochar added in the composting process, which improves the oxygen distribution in the matrix, making the aerobic microbiological processes more efficient (Sánchez-García et al. 2015).

Other researchers (Li et al. 2015; Malinska et al. 2014) have also emphasised an indirect benefit of biochar addition linked to the fact that biochar promotes microbial enzymatic activity that accelerates the degradation of organic matter. This specific outcome depends on the fact that biochar can adsorb molecules, such as NH_3, NH_4^+, H_2S and SO_4^2, that slow down the organic matter degradation rate when present in excess as free ions in the composting matrix. Jindo et al. (2016) also observed that the addition of biochar to the composting process has a direct influence on the microbial community, by increasing bacterial biomass over fungal biomass over time. However, the study did not indicate any benefit, though the capacity to shift the microbial population might have attractive applications when dealing with difficult organic waste having a high organic load.

Biochar not only accelerates the degradation process, but also increases the stability of the organic matter in the soils by increasing the aromatic character of the mixture (Zhang et al. 2014). The addition of biochar to the composting matrix protects the humic substance against decay by adsorbing them (Jindo et al. 2016). Furthermore, biochar in the composting matrix improves the organic matter humification, especially reducing the fulvic acids fraction, which are recognised as the least stable.

An additional interesting characteristic of biochar is the ability to adsorb heavy metals (Inyang et al. 2016) and organic contaminants commonly found in sewage sludge. Many studies have confirmed the ability of biochar to reduce the availability of heavy metals in the compost from sewage sludge (Cui et al. 2016; Borchard et al. 2012; Chen et al. 2010). Sorption of heavy metals by biochar involves several mechanisms, which mainly depend on the structure of the biochar itself. This is probably also the reason why the feedstock for the biochar and its production method are major determinants affecting the bioavailability of heavy metals. Some bioavailability studies have documented divergent results. For example, Lopez-Cano et al. (2016) found that the addition of biochar to composting had no effects on the bioavailability of heavy metals. Related to land application, the effect of ageing of BCM on the bioavailability of contaminants is still unknown. In addition, only limited information is available on the effect of biochar on organic contaminants, and the underlying mechanisms. Oleszczuk et al. (2014) reported that the bioavailability of polycyclic aromatic hydrocarbons (PAH) was reduced in biochar-added composting sewage sludge. In the case of organic contaminants, biochar might also stimulate the development of microorganisms that can degrade them (Godlewska et al. 2017). However, more research is needed to identify the mechanisms of sorption and to understand how to increase the microbial population responsible for the degradation of the organic contaminants.

3.3 Artificially Formed Terra Preta and Crop Growth

The potential positive synergy between biochar and the composting process has been confirmed also at a practical level, though the number of studies on this topic is still limited. In particular, very interesting results have been reported in cases where a BCM has been applied on land to increase crop yield.

The application of BCM on ferralsols in Australia has increased the maize yield and plant growth in terms of grain production and leaves chlorophyll content and nutrients uptake (Agegnehu et al. 2016a). Results from the land application of BCM on Ethiopian fields cropped to barley also showed similar results (Agegnehu et al. 2016b), and paralleled earlier results on oat cultivated either in a greenhouse or in open fields in the tropics (Schulz et al. 2013). The BCM proved to be effective in terms of crop yield when compared to a separate, single application of biochar or compost, or a chemical fertiliser. Nevertheless, the research was not yet able to highlight the mechanisms behind the interaction between biochar and compost in the enhancement of plant growth. Another study revealed the potential of BCM to counteract stress caused by water deficit in the cultivation of cucumber in a controlled greenhouse environment (Nadeem et al. 2017). At the same time, the study confirmed the capacity of the mixture to improve plant growth and yield.

The results from trials conducted in temperate climate zones are not as promising as those realised in tropical areas. Schmidt et al. (2014) tested the use of BCM in a vineyard soil with low fertility over a period of 3 years. The application of the mixture did not have a positive effect on any of the parameters analysed and showed only a minor positive effect on plant growth of the first year that disappeared in time. The results from another study (Seehausen et al. 2017) revealed that the application of the mixture even had a negative effect on plant growth and the total leaf area of the plants. This research was conducted in a controlled environment and only with composted spent substrate from the cultivation of mushrooms. Therefore, the results are applicable only to a very limited selection of raw materials and cannot be generalised.

4 Untapped Economic Potential of Organic Waste

The simplification of the criteria to assess sustainability has also justified the implementation of costs analysis schemes that do not take into account economic variables that cannot be directly connected to investments or operational costs. The lack of comprehensive financial analysis has generated optimistic expectations that often do not match the real market situations. As a consequence, revenue schemes based on a linear technology approach are very sensitive to the variation of the gate fees generated from the different raw materials and their quality. This issue has been extensively studied: for example, in the case of biogas production technology (Boulamanti et al. 2013). In addition to the variability of gate fees, it must be noted that organic waste treatment plants are generally developed on a very local basis, while the products usually follow the prices of the global market. Therefore, the low output quantity cannot influence the demand and the corresponding market price (Philp et al. 2017). For these reasons, in a linear approach, a frequent practice is to compensate for the variability of the main sources of income by increasing the marketability of the by-products for applications permitted by environmental legislation, but which were not considered in the economic models. However, the sustainability of practices associated with the sale of by-products needs further investigation. As an example, in the economy of biogas plants, the sale of digestate, the main by-product, plays an important role in the overall economic profitability of the plant (Gebrezgabher et al. 2010). As a consequence, digestate is often marketed as a very efficient organic fertiliser, despite the fact that the sustainability of this practice is debatable as the benefit in terms of nutrients recycling efficiency is not evident. In fact, the characteristics of digestate might promote the formation of nitrous oxides, a potent GHG when applied to the soil (Nkoa 2013). Recent field studies have confirmed the hypothesis that the land application of digestate can increase the nitrous oxide emissions to above normal levels (Eickenscheidt et al. 2014; Fiedler et al. 2017). It must be noted that in several countries or regions, digestate application to soil is not even practicable. Similarly, composting technology in a linear approach is very dependent on gate fees and the sale of the end product. The presence of municipal sewage sludge, which is a feedstock that has a high gate fee, limits the use of the end product from composting due to negative public perceptions. On the other hand, the uncontrolled use of other feedstocks to improve the appeal of the end product can increase the GHG emissions, thereby weakening the sustainability of the process (Lim et al. 2016a, b). By contrast, the use of pyrolysis technology to produce biochar in a linear approach would appear to be economically viable and sustainable at the same time, although the economic viability is currently questionable since this technology needs to be improved and scaled up, particularly for feedstocks that are not derived from lignocellulosic biomass (Roy and Dias 2017). In addition, recent studies do not support the conclusion of a reduced carbon footprint related to biochar, and the current biogenic carbon retention models might lead to incomplete conclusions in terms of sustainability (Guest et al. 2013).

The shortcomings of the linear approach associated with the latest developments in the sector of organic waste treatment have led to recent calls for a more diverse use of organic waste in order to fully exploit this multipurpose resource. New treatment techniques have gained more attention as a means to diversify the use of organic waste, secure more stable revenues in the long term and to improve the sustainability of the treatment processes and the corresponding production chains. Among these techniques, extraction of acids from the biogas process, production of animal feed by breeding insects on organic waste substrate and cultivation of algae in wastewater are among the most promising. Despite their higher profitability, the reason for their minor application in comparison to the generation of biogas or compost products depends on the fact that producing these alternative end products normally requires a specific feedstock, and a deeper understanding of the process, in order to be viable.

5 Barriers to the New, Holistic and Circular Approach

In spite of the positive effect that a circular holistic approach could induce, its application is still limited to experimental studies or small-scale projects, as there are several technical, economic and legislative issues that still prevent the adoption of such a model on a larger scale.

5.1 Lack of Innovative Supply Chains

The development of a circular holistic approach, and the creation of synergies and mutual benefits between the different sectors involved, requires in many cases the establishment of new supply chains for the new products. The products from these recycling processes are not easily integrated into the existing production chains. They frequently require extensive evaluation before substituting the traditional raw materials in the industry without compromising the main production process itself. The application of new raw materials is thus very slow in the food processing industry and in agriculture sectors, which already operate in a very competitive environment. An additional hindrance is the prevailing reluctance to accept anything generated from organic waste. This reluctance arises from a lack of knowledge of the long-term consequences related to their use.

5.2 Subsidising Policies

However, the most significant factors slowing down the development of a holistic approach are governmental subsidies policies. While the aim of supporting policies is to boost the green economy, evidence shows that in most cases they work as a deterrent to innovation and the development of new markets (Carus et al. 2014). As a consequence, organic waste resources are not properly allocated and most of the potential is lost due to erroneous intervention. In fact, current policies that support the production of biofuels and bioenergy from organic waste effectively prevent the development of and investment in products with a higher market value, such as proteins or natural acids. These policies are mainly justified by the effort to reduce GHG emissions and secure energy supply (Keegan et al. 2013). However, they fail to address the importance of other organic waste-based products with a higher market value according to the cascading concept as they do not provide any mechanism that would promote their application on a wide basis. The contradictory outcome of policies supporting the use of organic waste as a source of energy is also evident in other aspects. Usually, these policies have a secondary target, namely the development of rural areas by supporting agriculture through the generation of alternative revenue streams. However, the outputs (biofuels and bioenergy) of the new agricultural activities are not so significant that it can influence the global market. At the same time, the price fluctuations of fossil fuels are often a challenge for the economics of the production on a rural level (Keegan et al. 2013). This creates a further increase of the problems in rural areas without capitalising on the potential of all organic waste available in agricultural activities.

5.3 Non-holistic Policies

Governmental policies have also contributed to the wide adoption of the linear approach, though that was probably not the intended objective. As a matter of fact, many governmental institutions have issued specific policies to improve sustainability practices, where the main focus is on the reduction of GHG emission. This, in turn, has favoured the application of processes where the benefit gained from the reduced carbon footprint is easily comparable against the emissions from the fossil raw material. The financial support provided by governments has then led to a situation where most often the only criterion applied to measure sustainability is the capability to reduce GHG emissions (Bosch et al. 2015). Further benefits for the circular economy may come from the harmonisation of environmental and agricultural legislation.

6   Looking Ahead to the Future Development of BCM

BCM clearly has the potential to offset the negative consequences of agriculture intensification and the larger amount of waste from the food industry by enhancing the management of nutrient recycling and providing stable soil organic matter. In addition, BCM fits seamlessly into the bioeconomy concept as it can utilise, through the holistic approach, by-products and materials that are formed in other organic waste treatment processes but that are not suitable as such for processing with the available treatment options. The flexibility in terms of input materials also makes it possible to solve any problems with feedstock availability by processing waste from different sectors, such as the food industry, agriculture or wastewater treatment. However, before the wide application of the circular holistic approach can take place, there are still some challenges to be solved. In addition to a lack of policies that promote nutrients recycling and the restoration of organic matter in the soil, there are still technical and environmental aspects that must be addressed in order to understand the impact of BCM on the ecosystem. As discussed above, several studies dealing with BCM have been conducted, but these studies lack a common and systematic approach that would make it possible to predict the precise quality of the end product based on the characteristics of the feedstock. Currently, the experimental results seem promising only for a few raw materials; they cannot be generalised.

In strictly operational terms, future studies should focus on the effect of biochar addition on the biomass prior to the composting process. Though the results from previous studies are not always consistent, this set-up seems to be the most promising since it allows us to study not only the benefits of the combination of biochar and compost but also the mechanisms that regulate the interactions of these materials during the composting process. As described above, the addition of biochar to the compost process improves the porosity of the material and increases the temperature (Li et al. 2015; Theeba et al. 2012; Zhang et al. 2014). Conversely, it is likely that the modification of the operational parameters of composting, such as moisture or air feeding rate, affects the characteristics of the end products by changing the interactions between biochar and biomass during composting. Therefore, future studies should aim to identify how to adjust the process parameters in order to achieve the desired output irrespective of the feedstock.

The possibility to further research on the mechanisms involved in the interaction between biochar and compost will also help specify more accurately the characteristics of the mixture and narrow down the applications. At present, the different combinations of biochar and compost are usually applied indistinctively as soil amendments, organic fertilisers, and as a means to reduce persistent organic pollutants, without any assessment of the quality of feedstock or the contribution of process parameters. Despite the flexibility in being able to treat several types of organic waste, it is unlikely that one single combination of feedstock and process parameters can result in an end product that is optimal for all different applications. Future strategies should therefore focus on selecting suitable applications based on the local requirements and tailoring the process conditions in composting to produce end products with the desired characteristics. In this context, the characterisation of the feedstock emerges as an important factor in the optimisation of the efficiency in the selected application. In fact, the final use of the biochar and compost mixture restricts the selection of feedstock, as not all organic waste will be completely suitable for the selected process in terms of quality. As a consequence, mainly feedstock with the most suitable features will be processed. Unsuitable organic waste streams will either be processed to produce BCMs with different characteristics, or simply redirected to a more suitable treatment process within the holistic circular model for managing organic waste. This will further contribute to the increase of the overall material efficiency and streamline the management of organic waste streams towards sustainable development.

The technological developments need to be accompanied by novel approaches to measuring the degree of sustainability of BCM and its use. As one of the targets of BCM is to properly manage the nutrients cycle, it is logical to introduce new means to evaluate the potential to reuse nutrients. There are some techniques available to analyse the flow of nutrients in the natural cycle. While these techniques lack the ability to evaluate the efficiency of the recycling processes in qualitative terms, Grönman et al. (2016) proposed an approach that would take into account the efficiency of reusing nutrients in relation to their available amount. This novel method to calculate the 'nutrient footprint' offers the potential to help understand which processes and applications can increase the efficiency of reusing nutrients in various production chains. The degree of sustainability of the BCM can be further studied by integrating life cycle assessment and risk analysis. Life cycle assessment and risk analysis focus on other environmental aspects, such as carbon sequestration, eutrophication, soil acidification and contamination, associated with the recycling and reuse of organic waste in agriculture, which is both the initial generator and the final user of the end products in the holistic, circular approach to managing organic waste (Oldfield et al. 2018). In particular, the integration of parameters (e.g. carbon sequestration, eutrophication and soil acidification) that indicate the health of soil is of paramount importance to ensure sustainable development (Oldfield et al. 2018). As a matter of fact, soil depletion is one of the major threats to sustainable agriculture (Pimentel et al. 1995), but a circular holistic approach, by proper management of nutrients and constant replacement of organic matter in the top soil, can reverse the current adverse trends. It must be noted, however, that new means to assess sustainability require further validation before an extensive adoption. In fact, these new tools have been tested on a very limited range of specific conditions and only with one single crop or food chain. It is also clear that the sustainability of the circular holistic approach should be supported by economic and social considerations. However, the integration of such aspects might be difficult at a general level.

7 Conclusions

The circular holistic approach based on composting as a core treatment method offers a new model to cope with the mounting food demand by exploiting in full the potential of organic waste resources and, consequently, reducing the burden on the environment by combining existing and new treatment technologies. In particular, the synergies between biochar and compost offers the opportunity to improve the sustainability of the management of organic waste through efficient reuse of nutrients and return of the organic matter back to the soil. This new approach needs to be further supported not only by a deeper understanding of the mechanisms involved that cause the positive synergism of biochar and compost, but also by appropriate methods for sustainability assessment and encouraging policies. Common guidelines for the production of mixtures of biochar and compost should be developed in parallel with instruments that measure the sustainability with a broad scope. These new instruments should include tools to evaluate not only the reduction in GHG emissions but also other environmental parameters, the focus being on the efficiency in recycling nutrients and carbon sequestration of soil. Correspondingly, the possibility to quantify the sustainability of the new approach presented in this chapter will make it possible to revise the current environmental policies in order to promote sustainability as a whole, and not only from a narrow viewpoint.

References

Agegnehu, G., Bass, A. M., Nelson, P. N., & Bird, M. (2016a). Benefits of biochar, compost and biochar/compost for soil quality, maize yield and greenhouse gas emissions in a tropical agricultural soil. *Science of the Total Environment, 543*, 295–306.

Agegnehu, G., Nelson, P. N., & Bird, M. I. (2016b). The effects of biochar, compost and their mixture and nitrogen fertilizer on yield and nitrogen use efficiency of barley grown on a Nitisol in the highlands of Ethiopia. *Science of the Total Environment, 569–570*, 869–879.

Awasthi, M. K., Wang, Q., Huang, H., Li, R., Shen, F., Lahori, A. H., Wang, P., Guo, D., Guo, Z., Jiang, S., & Zhang, Z. (2016a). Effect of biochar amendment on greenhouse gas emission and bio-availability of heavy metals during sewage sludge co-composting. *Journal of Clean Production, 135*, 829–835.

Awasthi, M. K., Wang, Q., Ren, X., Zhao, J., Huang, H., Awasthi, S. K., Lahori, A. H., Li, R., Zhou, L., & Zhang, Z. (2016b). Role of biochar amendment in mitigation of nitrogen loss and greenhouse gas emission during sewage sludge composting. *Bioresource Technology, 219*, 270–280.

Borchard, N., Prost, K., Kautz, T., Moeller, A., & Siemens, J. (2012). Sorption of copper (II) and sulphate to different biochars before and after composting with farmyard manure. *European Journal of Soil Science, 63*, 399–409.

Bosch, R., Van De Pol, M., & Philp, J. (2015). Define biomass sustainability. *Nature, 523*, 526–527.

Boulamanti, A. K., Maglio, S. D., Giuntoli, J., & Agostini, A. (2013). Influence of different practices on biogas sustainability. *Biomass and Bioenergy, 53*, 149–161.

Buckwell, A., & Nadeu, E. (2016). *Nutrient recover and reuse (NRR) in European agriculture. A review of the issues, opportunities, and actions.* Brussels: RISE Foundation.

Carus, M., Dammer, L., Hermann, A., & Essel, R. (2014). *Proposals for a reform of the Renewable Energy Directive to a Renewable Energy and Materials Directive (REMD). Going to the next level: Integration of bio-based chemicals and materials in the incentive scheme.* Nova paper on biobased economy 2014–05, Nova-Institüt, Huerth, Germany.

Chen, Y. X., Huang, X. D., Han, Z. Y., Huang, X., Hu, B., Shi, D. Z., & Wu, W. X. (2010). Effects of bamboo charcoal and bamboo vinegar on nitrogen conservation and heavy metals immobility during pig manure composting. *Chemosphere, 78*, 1177–1181.

Cui, E., Wu, Y., Zuo, Y., & Chen, H. (2016). Effect of different biochars on antibiotic resistance genes and bacterial community during chicken manure composting. *Bioresource Technology, 203*, 11–17.

Eickenscheidt, T., Freibauer, A., Heinichen, J., Augustin, J., & Drösler, M. (2014). Short-term effects of biogas digestate and cattle slurry application on greenhouse gas emissions affected by N availability from grasslands on drained fen peatlands and associated organic soils. *Biogeosciences, 11*, 6187–6207.

Ellen Mac Arthur Foundation. (2019). *Approach: We inspire and enable the transition to a circular economy.* https://www.ellenmacarthurfoundation.org/our-work/approach. Accessed June 2019.

European Commission. (2012). *Innovating for sustainable growth: A bioeconomy for Europe.* Communication from the Commission to the European Parliament, the Council, the European Economic and Social Committee and the Committees of the Regions, European Commission, B-1049 Brussels, Belgium.

European Commission. (2018). *A sustainable bioeconomy for Europe: Strengthening the connection between economy, society and the environment: Updated Bioeconomy Strategy.* Directorate-General for Research and Innovation Unit F, RTD BIOECONOMY COMMUNICATION, European Commission, B-1049 Brussels, Belgium. pp. 107, PDF. https://doi.org/10.2777/792130. https://ec.europa.eu/research/bioeconomy/pdf/ec_bioeconomy_strategy_2018.pdf#view=fit&pagemode=none. Viewed June 2019.

FAO and RUAF Foundation. (2015). *A vision for city region food systems – building sustainable and resilient city regions.*

Fiedler, S. R., Augustin, J., Wrage-Mönnig, N., Jurasinski, G., Gusovius, B., & Glatzel, S. (2017). Potential short-term losses of N2O and N2 from high concentrations of biogas digestate in arable soils. *The Soil, 3*, 161–176.

Fischer, D., & Glaser, B. (2012). Synergisms between compost and biochar for sustainable soil amelioration. In S. Kumar & A. Barthi (Eds.), *Management of organic waste* (pp. 167–199). Rijeka: InTech.

Foley, J. A., Ramankutty, N., Brauman, K. A., Cassidy, E. S., Gerber, J. S., Johnston, M., Mueller, N. D., O'Connell, C., Ray, D. K., West, P. C., Balzer, C., Bennett, E. M., Carpenter, S. R., Hill, J., Monfreda, C., Polasky, S., Rockström, J., Sheehan, J., Siebert, S., & Tilman, D. (2011). Solutions for a cultivated planet. *Nature, 478*, 337–342.

Gebrezgabher, S. A., Meuwissen, M. P., Prins, B. A., & Lansink, A. G. O. (2010). Economic analysis of anaerobic digestion – A case of green power biogas plant in The Netherlands. *NJAS-Wageningen Journal of Life Sciences, 57*(2), 109–115.

Glaser, B., & Birk, J. J. (2012). State of the scientific knowledge on properties and genesis of anthropogenic dark earths in Central Amazonia [terra preta de indio]. *Geochimica et Cosmochimica Acta, 82*, 39–51.

Godlewska, P., Schmidt, H. P., Ok, Y. S., & Oleszczuk, P. (2017). Biochar for composting improvement and contaminants reduction. A review. *Bioresource Technology, 246*, 193–202.

Grönman, K., Ypyä, J., Virtanen, Y., Kurppa, S., Soukka, R., Seuri, P., Finér, A., & Linnanen, L. (2016). Nutrient footprint as a tool to evaluate the nutrient balance of a food chain. *Journal of Cleaner Production, 112*, 2429–2440.

Guest, G., Bright, R. M., Cherubini, F., & Strømman, A. H. (2013). Consistent quantification of climate impacts due to biogenic carbon storage across a range of bio-product systems. *Environmental Impact Assessment Reviews, 43*, 21–30.

Gul, S., Whalen, J. K., Thomas, B. W., Sachdeva, V., & Deng, H. (2015). Physico-chemical properties and microbial responses in biochar-amended soils: Mechanisms and future directions. *Agriculture Ecosystems and Environment, 206*, 46–59.

Hua, L., Wu, W., Liu, Y., McBride, M. B., & Chen, Y. (2008). Reduction of nitrogen loss and Cu and Zn mobility during sludge composting with bamboo charcoal amendment. *Environmental Science and Pollution Research, 16*, 1–9.

Inyang, M. I., Gao, B., Yao, Y., Xue, Y., Zimmerman, A., Mosa, A., Pullammanappallil, P., Ok, Y. S., & Cao, X. (2016). A review of biochar as a low-cost adsorbent for aqueous heavy metal removal. *Critical Reviews in Environmental Science and Technology, 46*, 406–433.

Jindo, K., Sonoki, T., Matsumoto, K., Canellas, L., Roig, A., & Sanchez-Monedero, M. A. (2016). Influence of biochar addition on the humic substances of composting manures. *Waste Management, 49*, 545–552.

Keegan, D., Kretschmer, B., Elbersen, B., & Panoutsou, C. (2013). Cascading use: A systematic approach to biomass beyond the energy sector. *Biofuels, Bioproducts and Biorefining, 7*, 193–206.

Kirschenmann, F. (2010). Alternative agriculture in an energy- and resource-depleting future. *Renewable Agriculture and Food Systems, 25*, 85–89.

Laird, D. A. (2008). The charcoal vision: A win-win-win scenario for simultaneously producing bioenergy, permanently sequestering carbon, while improving soil and water quality. *Agronomy Journal, 100*, 178–181.

Li, R., Wang, Q., Zhang, Z., Zhang, G., Li, Z., Wang, L., & Zheng, J. (2015). Nutrient transformation during aerobic composting of pig manure with biochar prepared at different temperatures. *Environmental Technology, 36*, 815–826.

Lim, S. L., Lee, L. H., & Wu, T. Y. (2016a). Sustainability of using composting and vermicomposting technologies for organic solid waste biotransformation: Recent overview, greenhouse gases emissions and economic analysis. *Journal of Clean Production, 111*, 262–278.

Lim, S. L., Lee, L. H., & Wu, T. Y. (2016b). Sustainability of using composting and vermicomposting technologies for organic solid waste biotransformation: Recent overview, greenhouse gases emissions and economic analysis. *Journal Clean Production, 111*, 262–278.

López-Cano, I., Roig, A., Cayuela, M. L., Alburquerque, J. A., & Sánchez-Monedero, M. A. (2016). Biochar improves N cycling during composting of olive mill wastes and sheep manure. *Waste Management, 49*, 553–559.

Malinska, K., Zabochnicka-Świątek, M., & Dach, J. (2014). Effects of biochar amendment on ammonia emission during composting of sewage sludge. *Ecological Engineering, 71*, 474–478.

Nadeem, S., Imran, M., Naveed, M., Khan, M., Ahmad, M., Zahird, Z., & Crowleyb, D. (2017). Synergistic use of biochar, compost and plant growth-promoting rhizobacteria for enhancing cucumber growth under water deficit conditions. *Journal of the Science of Food and Agriculture, 97*, 5139–5145.

Nkoa, R. (2013). Agricultural benefits and environmental risks of soil fertilization with anaerobic digestates: A review. *Agronomy for Sustainable Development, 34*, 473–492.

Oldfield, T., Sikirica, N., Mondini, C., Lopez, G., Kuikman, P., & Holden, N. (2018). Biochar, compost and biochar-compost blend as options to recover nutrients and sequester carbon. *Journal of Environmental Management, 218*, 465–476.

Oleszczuk, P., Zielińska, A., & Cornelissen, G. (2014). Stabilization of sewage sludge by different biochars towards reducing freely dissolved polycyclic aromatic hydrocarbons (PAHs) content. *Bioresource Technology, 156*, 139–145.

Philp, J., Schieb, A., & Chelly, M. (2017). Understanding value chains in industrial biotechnology. *Annales des Mines – Réalités industrielles, 1*, 56–65.

Pimentel, D., Harvey, C., Resosudarmo, P., Sinclair, K., Kurz, D., Mcnair, M., Crist, S., Shpritz, L., Fitton, L., Saffouri, R., & Blair, R. (1995). Environmental and economic costs of soil erosion and conservation benefits. *Science, 267*, 1117–1123.

Roy, P., & Dias, G. (2017). Prospects for pyrolysis technologies in the bioenergy sector: A review. *Renewable and Sustainable Energy Review, 77*, 59–69.

Sánchez-García, M., Alburquerque, J. A., Sánchez-Monedero, M. A., Roig, A., & Cayuela, M. L. (2015). Biochar accelerates organic matter degradation and enhances N mineralization during composting of poultry manure without a relevant impact on gas emissions. *Bioresource Technology, 192*, 272–279.

Schmidt, H. P., Kammann, C., Niggli, C., Evangelou, M. W. H., Mackie, K. A., & Abiven, S. (2014). Biochar and biochar-compost as soil amendments to a vineyard soil: Influences on plant growth, nutrient uptake, plant health and grape quality. *Agriculture, Ecosystems and Environment, 15*, 117–123.

Schulz, H., Dunst, G., & Glaser, B. (2013). Positive effects of composted biochar on plant growth and soil fertility. *Agronomy for Sustainable Development, 33*, 814–827.

Seehausen, M. L., Gale, N. V., Dranga, S., Hudson, V., Liu, N., Michener, J., Thurston, E., Williams, C., Smith, S. M., & Thomas, S. C. (2017). Is there a positive synergistic effect of biochar and compost soil amendments on plant growth and physiological performance? *Agronomy, 7*.

Sutton, M. A., Bleeker, A., Howard, C. M., Bekunda, M., Grizzetti, B., de Vries, W., van Grinsven, H. J. M., Abrol, Y. P., Adhya, T. K., Billen, G., Davidson, E. A, Datta, A., Diaz, R., Erisman, J. W., Liu, X. J., Oenema, O., Palm, C., Raghuram, N., Reis, S., Scholz, R. W., Sims, T., Westhoek, H., and Zhang, F. S., with contributions from Ayyappan, S., Bouwman, A. F., Bustamante, M., Fowler, D., Galloway, J. N., Gavito, M. E., Garnier, J., Greenwood, S., Hellums, D. T., Holland, M., Hoysall, C., Jaramillo, V. J., Klimont, Z., Ometto, J. P., Pathak, H., Plocq Fichelet, V., Powlson, D., Ramakrishna, K., Roy, A., Sanders, K., Sharma, C., Singh, B., Singh, U., Yan, X. Y., & Zhang, Y. (2013). *Our Nutrient World: The challenge to produce more food and energy with less pollution*. Global Overview of nutrient management. Centre for Ecology and Hydrology, Edinburgh on behalf of the Global Partnership on Nutrient Management and the International Nitrogen Initiative.

Tammeorg, P., Simojoki, A., Mäkelä, P., Stoddard, F. L., Alakukku, L., & ja Helenius, J. (2014). Biochar application to a fertile sandy clay loam in boreal conditions: Effects on soil properties and yield formation of wheat, turnip rape and faba bean. *Plant and Soil, 374*(1–2), 89–107.

The Waste Framework Directive. (2008/98/EC). (2008). On waste and repealing certain Directives. *Official Journal of the European Union*, L 312/3.

The White House. (2012). National Bioeconomy Blueprint, 44.

Theeba, M., Bachmann, R. T., Illani, Z. I., Zulkefli, M., Husni, M. H. A., Samsuri, A. W., & Samsuri, W. (2012). Characterization of local mill rice husk charcoal and its effect on compost properties. *Malaysian Journal of Soil Science, 16*, 89–102.

United Nations. (2018). *About the sustainable development goals*. https://www.un.org/sustainabledevelopment/sustainable-development-goals. Accessed June 2019.

Vandecasteele, B., Sinicco, T., D'Hose, T., Vanden Nest, T., & Mondini, C. (2016). Biochar amendment before or after composting affects compost quality and N losses, but not P plant uptake. *Journal of Environmental Management, 168*, 200–209.

Vandecasteele, B., Willekens, K., Steel, H., D'Hose, T., Waes, C., & Bert, W. (2017). Feedstock mixture composition as key factor for C/P ratio and phosphorus availability in composts: Role of biodegradation potential, biochar amendment and calcium content. *Waste and Biomass Valorization, 8*, 2553–2567.

Viaene, J., Van Lancker, J., Vandecasteele, B., Willekens, K., Bijttebier, J., Ruysschaert, G., De Neve, S., & Reubens, B. (2016). Opportunities and barriers to on-farm composting and compost application: A case study from northwestern Europe. *Waste Management, 48*, 181–192.

Webster, P., & Francis, R. (2016). Green shoots of success. Drivers and approaches towards a successful bioeconomy. *The Chemical Engineering, 900*, 24–28.

Xu, G., Wei, L. L., Sun, J. N., Shao, H. B., & Chang, S. X. (2013). What is more important for enhancing nutrient bioavailability with biochar application into a sandy soil: Direct or indirect mechanism? *Ecological Engineering, 52*, 119–124.

Zhang, J., Lü, F., Shao, L., & He, P. (2014). The use of biochar-amended composting to improve the humification and degradation of sewage sludge. *Bioresource Technology, 168*, 252–258.

Composting: A Sustainable Route for Processing of Biodegradable Waste in India

Ashootosh Mandpe, Sweta Kumari, and Sunil Kumar

Abstract Surging populations, coupled with the ever-increasing demand for sustenance, have led to the generation of behemoth proportions of wastes throughout the globe. The processing of such a considerable amount of waste has raised concerns for environmental planners, policymakers, and researchers in regard to maintaining sustainability. Biodegradable waste is a part of the total waste stream. Consideration should be given to the importance of making better use of biodegradable waste. The technology that is adopted for the management of biodegradable waste should be ecologically sustainable and cost-effective, as well as beneficial to social well-being. The most efficient way of managing biodegradable waste must include different methods for the optimal utilisation of such waste, ranging from the small scale (single household) to the very large scale (entire city). Amid all the other waste processing technologies, composting stands out as a most potent option because of its ability to maintain and restore soil fertility, along with the transformation of waste into a resource. Composting is one of the few technologies which has a benefit–cost ratio higher than 1 at all scales of operation. This chapter analyses the most significant aspects of the composting process, including the recent developments and dynamics involved in it. The chapter discusses various aspects of composting via analysis of the integrated waste management system and composting-related projects implemented at the community level in the Indian context. Finally, the chapter presents policies and the efforts put in place by the Government of India with the aim of encouraging composting practice and related activities.

Keywords Biodegradable waste · Biological decomposition · Community-based implementation · Composting · Composting techniques · India · Municipal solid waste (MSW) management · Sustainable utilisation

A. Mandpe · S. Kumari · S. Kumar (✉)
Council of Scientific & Industrial Research (CSIR)-National Environmental Engineering and Research Institute (CSIR-NEERI), Nagpur, India

1 Introduction

The swelling population, along with ever-increasing demands, has led to the generation of behemoth volumes of waste. The waste generation rate is proportional to the human population throughout the globe. Management of these enormous volumes of generated waste tends to put pressure on urban local bodies governing the cities or the urban parts of the Indian country. In the twenty-first century, solid waste management has gained significant consideration from various environmental planners, policymakers, non-governmental organisations and waste managers because of the negative environmental impacts resulting from unscientific management and adoption of unsafe disposal practices in respect of waste. The situation is particularly distressing in developing countries: municipal solid waste (MSW) can be spotted almost ubiquitously in the urban as well as semi-urban parts of most developing countries. Eco-friendly MSW management has become a challenging task in India due to the increasing population, unparalleled and unalterable urbanisation, as well as industrial development (Ramachandra et al. 2018). To give an idea of the size of the problem, Table presents today's MSW generated per capita in India and other developing countries, as well as the forecast figures (Hoornweg and Bhada-Tata 2012).

MSW poses a considerable risk to the surrounding environment, including water, soil, and human health. According to the Central Pollution Control Board (CPCB) report, the MSW in India has significantly higher fraction of organics (CPCB 2013a). Enormous volumes of biodegradable solids get generated as earthly and aquatic weeds, leaf litter, and agricultural waste. The weeds, if left untreated, may infest the land and water resources and may lead to their depletion. In developing nations, the practice of burning agricultural waste and leaf litter in the open air is still prevalent, which not only abolishes the significant deal regarding carbon and other nutrients but also causes air pollution, resulting in global warming.

According to Tomić and Schneider (2017), different approaches have been adopted as alternatives to traditional methods of solid waste management. These include the reuse of MSW and biodegradables waste, which is done via thermal processing and bioprocessing techniques. Thermal processing techniques comprise incineration, pyrolysis, and gasification; however, these do not seem feasible due to the low calorific values (800–1000 kcal/kg for Indian MSW) of MSW generated in developing countries, high energy intensiveness, as well as the fact that they result in other environmental hazards. Bioprocessing techniques comprise anaerobic digestion and composting. Among all other waste processing technologies, composting seems to be a powerful option because of its ability to maintain and restore soil fertility, along with the transformation of waste into a resource. Composting is among the best known processes for the biological stabilisation of biodegradable waste and is one of the few technologies which has a benefit–cost ratio greater than 1 for all scales of operation.

Composting can be explained as the process of transforming the organic substrate into nutrient-rich manure through the medium of microbial communities. The

Table Population and MSW generated per capita per day and projections for 2025 in India and other developing countries in the same region

Sr. no.	Country	Current available data			2025			
		Total urban population	MSW generation (kg/capita/day)	Total MSW generation (tonnes/day)	Total population	Urban population	MSW generation (kg/capita/day)	Total MSW generation (tonnes/day)
1	Bangladesh	38,103,596	0.43	16,384	2,06,024,000	76,957,000	0.75	57,718
2	Bhutan	225,257	1.46	329	819,000	428,000	1.7	728
3	China	511,722,970	1.02	520,548	1,445,782,000	822,209,000	1.7	1,397,755
4	Fiji	339,328	2.10	712	905,000	557,000	2.1	1170
5	Hong-Kong	6,977,700	1.99	13,890	8,305,000	8,305,000	2	16,610
6	India	321,623,271	0.34	109,589	1,447,499,000	538,055,000	0.7	376,639
7	Indonesia	117,456,698	0.52	61,644	271,227,000	178,731,000	0.85	151,921
8	Malaysia	14,429,641	1.52	21,918	33,769,000	27,187,000	1.9	51,655
9	Maldives	70,816	2.48	175	411,000	233,000	2.2	513
10	Nepal	3,464,234	0.12	427	38,855,000	10,550,000	0.7	7385
11	Pakistan	60,038,941	0.84	50,438	224,956,000	104,042,000	1.05	109,244
12	Singapore	4,839,400	1.49	7205	5,104,000	5,104,000	1.8	9187
13	Sri Lanka	2,953,410	5.10	15,068	20,328,000	3,830,000	4	15,320
14	Thailand	22,453,143	1.76	39,452	68,803,000	29,063,000	1.95	56,673
15	Vietnam	24,001,081	1.46	35,068	106,357,000	40,505,000	1.8	72,909

Source: Hoornweg and Bhada-Tata (2012)

prerequisite of the composting process is the availability of substrate in an organic origin (Gajalakshmi and Abbasi 2008a). More precisely, composting signifies the process of the biodegradation of the mixture of organic substrate conducted by the populations of various microbial species in aerobic environments in the solid state.

The process of composting generally takes place in four phases, and they may occur concomitantly, rather than sequentially (Belyaeva and Haynes 2009). These four phases include:

  (i) The mesophilic phase
 (ii) The thermophilic phase
(iii) The cooling phase
(iv) The curing phase

During the initial phase (the mesophilic phase), which is also termed the decomposition phase, the bacterial community present in the mixture of organic substrate combines oxygen with carbon to produce energy and carbon dioxide. A portion of the energy is utilised by the microorganisms for growth and reproduction processes, and the rest is released as heat. In this stage, the oxidation of easily degradable organic matter takes place, along with the proliferation of mesophilic bacteria, resulting in a rise in temperature of the substrate to be composted. These mesophilic bacteria may include *E. coli* and other bacterial strains which are inhibited by the temperature as the process is taken over by the thermophilic bacteria in the transition range.

The next phase, i.e. the thermophilic phase, which is also termed the stabilisation phase, involves the mineralisation of slowly degradable molecules, along with complex process like humification of lignocellulosic compounds. In this stage, a rise in temperature can be observed, which lasts only for a few days. After the completion of a thermophilic phase, the available manure seems to be digested, but the bristlier materials will remain intact. This phase is the cooling phase, in which the mesophilic microbes, which were hurtled away during the thermophilic phase, take over the system and start digesting the more resilient organic constituents of the substrate. Fungi and other macroorganisms like earthworms also enter back into the system.

A long curing time acts as a safety net for the destruction of remaining pathogens. The immature compost can be detrimental to plants and may even result in the production of phytotoxins. These phytotoxins may deprive the soil of nitrogen and oxygen and can contain higher levels of organic acids (Vandergheynst 2009).

2   Composting of Different Wastes

Composting is a cost-effective and clean option for waste disposal. It also offers extra benefits such as reduction of greenhouse gas (GHG) emissions, space taken by the landfill sites, and groundwater and surface water contamination by waste. The reduction in GHG emission consequently helps in regard to the issue of global warming/climate change (Hubbe et al. 2010).

Composting is a fundamental feature of an integrated solid waste management strategy. For effective integrated solid waste management, the key approaches are reducing, reusing, recycling, and managing waste to ensure environmental and human health safety. Local situations and needs vary from region to region and so does the composting process. Composting is a part of organic recycling, which converts organic waste into a nutrient-rich soil conditioner and also results in a reduction of GHG emissions (Lou and Nair 2009). The composting of different waste substrates is discussed in brief in the following subsections.

2.1 Agricultural/Lignocellulosic Waste

To maintain sufficient and long-lasting humus, biodegradable wastes, such as sawdust, wood shavings, coir pith, pine needles, and dry fallen leaves, are mixed together (Gajalakshmi and Abbasi 2008). However, it is observed that lignin-rich plant materials do not disintegrate rapidly. To enhance the decomposition, the waste material is treated with lime. They are mixed in a ratio of 5 kg of lime per 1000 kg of waste to prepare a good compost from hard plants. Lime can be in the form of dry powder or semi-solid substance when mixed with water. Liming weakens the lignin structure and enhances the humification process in plant residues, thus improving the humus quality (Hubbe et al. 2010). As an alternative to lime, phosphate rock in a dry powder state can be mixed in a ratio of 20 kg per 1000 kg of organic waste; it contains phosphates and micronutrients, which make a compost that is rich in plant nutrients.

2.2 Sewage Sludge

Sludge from sewage can be mixed with carbon by-products from agricultural sources, such as straw, sawdust, or wood chips, to produce compost. Microorganisms and parasites, which can cause diseases, are killed by the heat generated from bacteria digesting both sewage sludge and plant material in the presence of oxygen. Proper air circulation is obtained by maintaining aerobic conditions under 10–15% oxygen and bulking agents like shredded tyres. Stiff materials separated from softer leaves and lawn clippings maintain better ventilation (Fang et al. 1999).

An insulating blanket of previously composted sludge is placed over aerated composting piles for uniform distribution of pathogen-killing temperatures. The composting mixture should have an initial moisture level of about 50%, but if the temperature is inadequate for pathogen reduction, then the compost moisture content rises above 60%. Hence, composting mixtures piled on concrete pads with built-in air ducts might improve air circulation (Fang et al. 1999). A blower drawing vacuum may be used to minimise odours. When the moisture reaches 70%, exhausting through filtering pile can be replaced. The accumulated liquid in the underdrain ducting may be returned to the sewage treatment plant. For better moisture control, roofing of composting pads is required.

After a small interval that is sufficient for pathogen reduction, undigested bulking agents are recovered for reuse. The optimum initial carbon-to-nitrogen ratio of a composting mixture is between 26:1 and 30:1, but the amount required to dilute concentrations of toxic chemicals in the sludge to acceptable levels for the intended compost use determines the composting ratio of agricultural by-products. Agricultural by-products have less toxicity; however, suburban grass clippings containing residual herbicide levels are detrimental to some agrarian purposes, and freshly composted wood by-products may contain phytotoxins that inhibit the germination of seedlings until detoxified by soil fungi (Jouraiphy et al. 2005).

2.3   MSW

Composting yard trimmings, agricultural wastes, and sewage sludge have increased the interest in composting of organic fractions from MSW. MSW contains materials that vary in size, moisture, and nutrient content, and the organic fractions can be mixed with varying degrees of non-compostable wastes and possibly hazardous constituents. Producing a market-savvy compost product requires a range of physical processing technologies, in addition to the biological process management (Varma and Kalamdhad 2013).

Modern MSW composting system consists of four tasks: collection, contaminant separation, sizing and mixing, and biological decomposition. The collection task is representative of the characteristics of the incoming waste, which largely determines the processing requirements of the remaining tasks. The separation task generates recyclable and rejects streams. Biological activity can be enhanced by increasing the surface area of the waste in size reduction, whereas mixing ensures the adequacy of nutrients, moisture, and oxygen in the material.

Evaluation of system design: Several criteria are important, like cost (capital, operations, and maintenance), market specifications for compost and recyclable by-products, and system flexibility to respond to a changing MSW feedstock. The economic analysis of a composting facility should evaluate options like landfilling or incineration, along with different ways of achieving the same goal. It should also include analysis of cost between source separation, centralised separation, contaminant removal, and market assessment for recyclable by-product streams and the compost product (Wei et al. 2017). A composting facility should consider the quality control and product, as it might be a problem if a particular technology produces a less recyclable product as compared to markets. A flexible facility will be able to adapt to changes in the regulatory environment, in market specifications, and in the waste stream. Hence, the MSW facility should be flexible, in order to operate in the long term.

2.4　Biomedical Waste

Biomedical waste needs to be pre-treated with 5% sodium hypochlorite (NaOCl) at the disposal site for safety measures (Dinesh et al. 2010). It should be subjected to an initial decomposition process by blending it with cow dung slurry and can be further treated using the vermicomposting technique. Various epigeal species of worms can be utilised for this purpose. Practising this methodology for treating biomedical wastes can make these worms more efficient in carrying out the decomposition process. Vermicomposting with proper handling of biomedical waste can be an energy-efficient, eco-friendly approach for reducing and recycling this hazardous waste (Dinesh et al. 2010).

3　Composting Techniques Used in India

The organic fraction comprises the significant portion of waste generated in developing countries like India, and due to improper waste management practices, only a minor portion of the total waste volume is processed. The composting technique has been divided into two categories: the first category comprises conventional composting techniques, while the other category includes novel composting techniques. Each of the composting techniques is explained in brief in the following section. Figure shows different composting techniques adopted for processing various types of biodegradable waste.

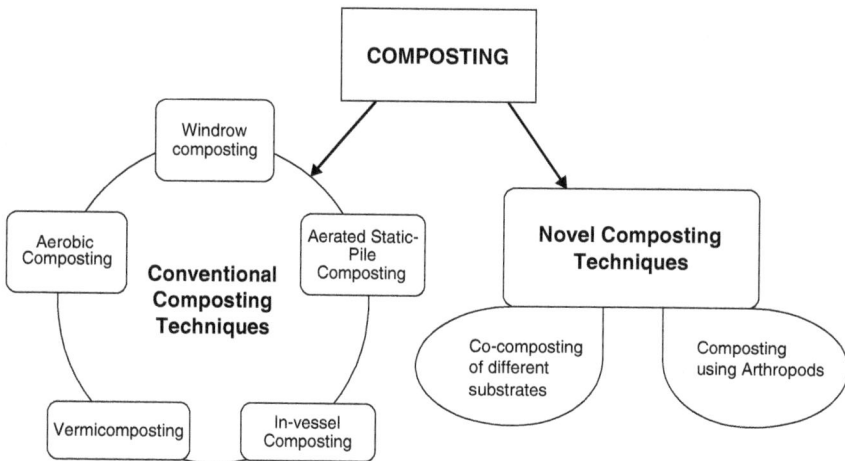

Fig.　Different composting techniques practised in India

3.1   Conventional Composting Techniques

Windrow Composting In a windrow composting system, piles of waste are kept for the decomposition and aeration activity and are turned simultaneously. The turning of waste piles leads to the lowering of the temperature of the compost and also allows the circulation of air in the system. Windrow systems are comparatively economical since there is no such requirement of any mechanical tools for providing aeration. The height of the windrow depends on the type of waste substrate and the equipment or tool used for flipping purposes. The intermittent flipping of the waste heaps is necessary for maintaining the optimal temperature (Kumar 2011).

Aerated Static Pile Composting In an aerated static pile composting system, the waste substrate is decomposed without any physical activity over 30–35 days. Perforated pipes are used for providing aeration to the waste piles. These waste piles can be kept open, covered, semi-covered, or in windrow form. The rate of aeration is one of the most critical factors in this type of composting system. The asperity of aeration produces the vertical temperature difference within the entire structure of the waste pile (Hubbe et al. 2010).

Vermicomposting The process of vermicomposting involves the utilisation of earthworms of different species for the disintegration of semi-decomposed biodegradable waste. It has been found that the earthworms can consume up to five times their body mass every day. According to CPCB (CPCB 2013b) report, the total number of composting and vermicomposting plants located in several states of India are shown in Fig.. Earthworm species commonly used in the vermicomposting systems are *Perionyx excavatus*, *Eudrilus eugeniae*, and *Eisenia foetida*. These epigeal species are extremely prolific feeders which can devour diverse streams of organic waste.

Aerobic Composting The aerobic composting system is carried out in the presence of air at a specific temperature and pH range where the microbial consortia decompose the biodegradable waste into a nutrient-rich bio-fertiliser (Monson and Murugappan 2010). During the aerobic composting process, the optimal rate of aeration can solve the problem of odours.

In-vessel Composting The in-vessel composting system involves the decomposition of the organic fraction of the waste in a closed container or vessel in a controlled environment (Manyapu et al. 2017). Various in-vessel composting systems comprise rotating drums and agitated bags. Instinctive aeration and mechanical agitation can enhance the decomposition process and boost the rapid composting process. Motorised agitation may lead to contravention of particles, providing a healthier contact of the microorganisms with the carbon, while the higher temperature in the vessels can finish the pathogenic activities in the system, and an unceasing

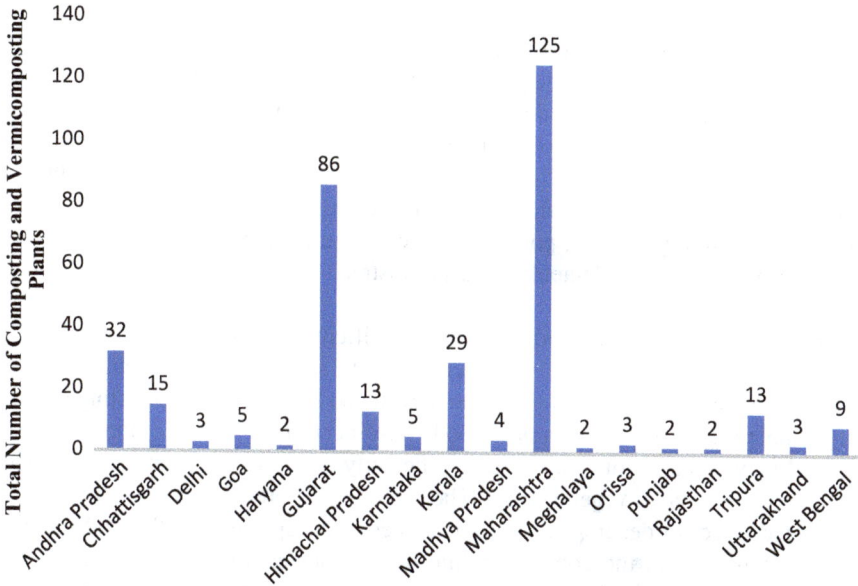

Fig. Total number of composting and vermicomposting plants across India. (Source: CPCB 2013b)

aeration results in the removal of odours from the system (Monson and Murugappan 2010).

3.2   Novel Composting Techniques

To overcome the problems associated with conventional composting practices and to improve the efficiency of the composting system, different researchers and environmentalists throughout the globe have adopted various approaches for effectual composting of biodegradable waste; these are termed novel composting techniques in this chapter. One of the approaches, called the co-composting technique, involves the utilisation of different waste substrates; another approach involves utilisation of arthropods like millipedes, black soldier flies, and similar organisms for composting purposes. Both of these techniques are explained in brief in this section.

Co-composting of Different Waste Substrates   This is a technique in which two different waste substrates are mixed together and applied to augment the composting activity. Co-composting of the waste substrate using specific type of bulking agent can offer sufficient particle structure and the denseness of the particles may tend to provide airspace needed for improving the aerobic microbial activity, leading to an increased rate of biodegradation. Various researchers have utilised food waste as a substrate for co-composting with other waste substrates because of their unique characteristics. Practising the co-composting of food waste can prove to be the best solution for processing food waste. Food waste can be blended with other waste substrates like chopped hay, rice husks, wooden chips, wheat straw, rice bran, sawdust, and other similar biodegradable waste to maintain the favourable carbon/ nitrogen (C/N) ratio, void spaces, moisture, and nitrogen content. The problem regarding the lower pH of the food waste can be solved by adding sodium acetate in the mixture of the waste, to balance the pH level of the system. Similarly, different feedstocks, like pig manure, poultry litter, etc., can be applied for co-composting activities with other supplementary waste substrates (Kumar et al. 2018).

Composting Using Arthropods Among the different techniques practised for the management of biodegradable waste, fertilising the soil is the most favoured one. In recent years, novel techniques have been developed for efficient composting of bio-degradable waste, and various arthropods have been used in the composting process. Many species of insects are efficiently utilised for reprocessing and transforming waste of vegetal origin. The blends of different types of waste formed at diverse stages of decomposition entice an essential species of arthropods whose life cycle is over in the compost, while any further addition to it enhances the decomposition process. These arthropodal species boost the development of the consentient species (Kumar et al. 2018). Arthropods like millipedes and black soldier fly larvae have been successfully used by Karthigeyan and Alagesan (2011) and Diener et al. (2015) for the composting of different types of organic waste. The millipedes transform the waste substrate into faecal pellets, which affect the required physicochemical characteristics by decreasing the C/N ratios (Karthigeyan and Alagesan 2011).

4   Composting as a Technique for Solid Waste Management in the Indian Scenario

Like other developing countries, India is also confronting solid waste management as one of the major issues it faces. Due to the expeditious growth in population, urbanisation, and industrialisation, India is switching towards an industrial and service-based economy and away from an agricultural-based economy. India has 29 states and 7 union territories and has climatic, geographical, ecological, cultural, traditional, and religious heterogeneity. The present urban population of India accounts for 31.2% (377 million) of its total population (Census 2011). As per CPCB (CPCB 2013b), India generated about 127,486 tonnes per day (TPD) of waste in 2011, out of which 89,334 TPD (70%) was collected; only 15,881 (12.45%) of the total waste was processed (CPCB 2013b). Being a developing country, the organic fraction of MSW in India is usually high. The urban MSW composition in Indian cities contains approximately 40% organic, 40% inert, and 19% of recyclable fractions (Sharholy et al. 2008). India has 279 composting plants, 138 vermicomposting facilities, 172 bio-methanation processors, 29 pelletisation plants, and 34 landfill sites across the country (Planning Commission 2014). From the numbers, it can be understood that composting technology is the most favoured and recommended among all the other technologies for efficient waste management. The problem of waste management is still prevalent in the country and is one of the major concerns for urban local bodies, which govern the urban parts of the country. In spite of the facilities above, it is believed that 90% of waste in India is dumped in an unscientific manner (Gajalakshmi and Abbasi 2008). The average per capita waste generated ranges from approximately 0.17 to 0.62 kg depending on location, economic activity, and culture (Kumar et al. 2009). The rapid composting technique is highly encouraged in big Indian cities. Indore composting centre is one of the foremost maintained composting plants in the country. Considering different zones in India, the following subsections discuss the current scenario of composting in several cities, along with the overall scenario of composting on the ground level.

4.1 Composting in the City of Kolkata

Kolkata city is the capital of the state of West Bengal. According to the South Asian Forum for the Environment, Kolkata generates 4000 tonnes of waste per day, out of which 37% is compostable (Dutta 2018). The major issue being faced by the Kolkata Municipal Corporation is the unavailability of land. The only landfill site in the region is "Dhapa", and it is already oversaturated from use over the last 30 years. The Dhapa dumpsite is located in the East Kolkata Wetlands, which is a Ramsar Convention Wetland. Dhapa dumpsite alone contributes to 7500 cubic metres of GHGs per hour (Times News Network 2013). According to a news daily, Kolkata city needs at least two solid waste treatment plants for the efficient management of waste (Dutta 2018). Keeping in view the massive problem of waste management, many housing complexes and societies have started installing compost machines within their premises for the treatment of MSW generated from household activities. The Belani Group of Companies, which is a housing development firm, has installed two composting facilities near the Hiland Riverside project in Batanagar. It has been announced that the Kolkata Municipal Corporation will be installing a similar composting facility in Tollygunge, Madhyamgram, and Rajarhat. These areas are among the most densely populated regions in Kolkata city and are also being developed at a rapid pace. Over 1100 apartments in South City (Residential Complex) in Kolkata have also joined this movement and have decided to sensitise the MSW for soon implementing the same mechanism in their housing complexes (Desai 2018). A small town in West Bengal named Uttarpara has helped Kolkata in receiving a prestigious award under the Urban Solid Waste Management category in the C40 Mayors' Summit held in Mexico on 1 December 2016, for its exclusive and effective solid waste management, which was mainly focussed on the segregation of waste at source. It was found that all the waste in Uttarpara is separated at source into biodegradable and non-biodegradable categories (Rao 2016). It has been reported that, earlier, a dumpsite, with debris ranging up to 50 ft. in length, was present in the town. However, a composting facility has been installed there and is operating successfully (Bhattacharya 2016). Nowadays, the composting facility at Uttarpara receives about 12–14 tonnes of raw biodegradable waste and produces 3 to 4 tonnes of manure per day (Rao 2016). The attitude of negligence towards the dumping site in Kolkata resulted in deplorable hygiene conditions and health impacts. Kolkata is a metropolitan city which is gradually moving towards a clean society. Small steps may gradually change the current scenario, but at the same time, the pace needs to be boosted to overcome the surging waste management problems.

Similarly, to promote sustainable waste management and utilise the nutrient-rich manure from composting processes, two female entrepreneurs from Kolkata, namely, Avantika Jalan and Rashmi Sarkar, have created a social enterprise called Mana Organics to recover soil fertility by producing chemical-free organic compost. Currently, they are operating four small composting projects: one in Tinsukia city, situated in the state of Assam, and the other three in villages in Madhya Pradesh. The compost produced from these composting facilities is marketed and sold in New Delhi and Kolkata. The firm started with a capital of 33.75 Lacs Rupees in 2011 (Basu 2014). Both entrepreneurs were inspired by their visit to Khargone district in the state of Madhya Pradesh. During their visit, they observed that a few regions in that district were devastated due to the failure of the cotton crop, and so they decided to help in solving the problem of the farmers (Basu 2014). The composting technique was adopted to manage the organic waste generated by agricultural and household activities to produce chemical-free organic manure, and they also spread awareness about the same among the farmers. Today, Mana Organics works with Adivasi (tribal) farmers in three villages of Madhya Pradesh: Sultanpur, Lachera, and Koriyakal (Basu 2014).

4.2   The Scenario of Composting in Delhi

Delhi is a union territory and the capital of India, and its present population is 11 million (Census 2011). According to a renowned environmental magazine, *Down to Earth*, Delhi generates 9500 TPD of waste, out of which 8000 TPD is collected and transported to three landfill sites, namely, Bhalswa, Ghazipur, and Okhla (Mudgal 2015). In some localities of the city, like Dwarka, Defence Colony, Sarita Vihar, Preet Vihar, and Dilshad Garden, composting facilities are already installed in their respective community parks, and segregation at source at the domestic level is being practised. The chairman of Delhi Municipal Corporation, Mr. Naresh Kumar, claimed that five localities in the state would become self-sustainable and would be transformed into a zero waste discharge site by the end of 2018 (Mudgal 2015). These five North Delhi Municipal localities are Pandara Park, Jor Bagh, Kaka Nagar, Bapa Nagar, and Golf Links. This idea was proposed in collaboration with a local resident welfare association. This plan emphasised segregation at source and separation into the wet waste, dry waste, and e-waste categories (Halder 2017). Delhi has three centralised composting plants at Narela, Bhalswa, and Okhla, which process around 500 TPD of biodegradable waste. Temples in Delhi are also coming forward to contribute to green practices. In Jhandewalan Temple, 5000–10,000 devotees visit every day, which results in the generation of 200 kgs of floral waste per day. On Tuesdays and Sundays, the waste generation rises to 500 kg, and during Navratri Puja (Goddess Festival), it goes up to about 1 TPD (Desai 2018).

In 2013 Delhi University invested 4 Lacs Rupees in the renowned college "Miranda House" to encourage composting within its premises to manage biodegradable waste most efficiently. The institute uses the raw material, comprising a mixture of horticulture and kitchen waste in a 1:3 ratio, for composting. The period required for composting is about 12–15 days (Mudgal 2015). The institute collaborated with Green Bandhu Environmental Solutions & Services, and now they produce 60 kgs of organic compost every day. It is encouraging the use of compost for recreational purposes in the region. The institute also earns around 4000–5000 Rupees by selling the supplementary compost produced from the composting plant and ultimately saves 12,000 Rupees which would have been incurred in the transport of waste (Mudgal 2015).

The General Pool Residential Complex in New Moti Bagh region is spread across 110 acres, with 1100 families. In 2013, this residential complex installed a composting facility provided by Green Planet Waste Management Pvt. Ltd. An area of 300 m² was needed for the composting facility, which receives around 700 kg of kitchen waste and 900 kg of horticultural waste per day (Mudgal 2015). The period required to get a matured compost is 15–20 days, thus producing organic compost. The company invested about 8 million Rupees in setting up this plant, and now the operators are struggling to recover the expense invested in installing the facility. A financial crisis or losses may result in its failure (Mudgal 2015).

The Defence Colony in the state of Delhi is practising composting for its kitchen waste using effective microorganisms (EM1) microbial solution-based pit composting. The composting period for this technology is relatively high, at 3–4 months. This setup requires a land area of just 30 m², costing 70,000 Rupees only (Mudgal 2015). The cost for every household served by this plant is 45 Rupees only, which is quite affordable. This setup was established by the Residential Welfare Association (RWA), with the help of a non-profit organisation named Toxics Link. Two rag pickers are trained by the RWA to operate the compost facility, and their wages are paid for from the compost facility itself.

Recently, the North Delhi Municipal Corporation has declared that it will shortly set up four "bio-methanation" and "aerobic drum compost plants" under its jurisdictions. The bio-methanation plants are proposed to deal with green waste, having an intake capacity of 5 TPD, and the compost plants will have a capacity of 1 TPD load. These plants will deal with green waste generated from RWAs, parks, and vegetable markets. The total expenditure for 5 years for the procurement, operation, and maintenance of the plants would be a minimum of 17.95 Crores (Adak 2018).

4.3   Status of Composting in Nagpur City

Nagpur is the third largest city in the state of Maharashtra and ranks 13th in terms of population in India. Nagpur is a central part of India, with a population of more than 2.4 million (Census 2011). The average waste collection in Nagpur is 1119 TPD, out of which only 200 TPD is processed by Hanjer Biotech Energies Pvt. Ltd. This company produces 35–40 tonnes of compost daily, which is sold to Rashtriya Chemicals and Fertilizers (Times News Network 2017). A waste characterisation of Nagpur city was performed by CSIR-NEERI in all ten zones of the city. The sampling sites were decided based on different sources of wastes, viz. residential (slum and non-slum areas) and institutional areas, community bins, commercial areas, and disposal site. The household waste composition indicated a 77% organic fraction, 11.60% plastic, 7.66% paper, and 3.74% textiles and cardboard. However, the average value of all ten zones had a waste composition comprising 60% organic fraction, 15.50% plastic, 11.20% paper, 2.10% inerts, and the remaining 11% including wood, metal, glass, and other parts (Dutta 2017). The total waste dumped during the year 2016–2017 was estimated at approximately around 14,000 TPD by the Nagpur Municipal Corporation (NMC). Initially, the garbage was processed for bio-mining and then segregation, harrowing, and finally spraying of bio-cultures to speed up the degradation. This process proved effective in reducing the height of the piled waste to some extent. However, there is a need to market the produced compost. A recent massive fire and the unpleasant odour of the waste indicate that enormous challenges remain (Times News Network 2017).

The NMC also launched one project in June 2017 to deal with garden waste. This was a pilot project: they planned to include five gardens in the city and a women's self-help group in the processing of biodegradable waste. This initiative was undertaken to reduce the burden of waste, in terms of transportation, as well as at the dumpsite. Provision was made to utilise the compost produced by any garden in the same garden (Dutta 2017).

4.4   The Scenario of Composting in Alappuzha and Thiruvananthapuram

Alappuzha and Thiruvananthapuram are cities located in the state of Kerala in the southern part of India. The populations of Alappuzha and Thiruvananthapuram 1,74,176 and 7,43,691, respectively (Census 2011). The Centre for Science and Environment, which is a not-for-profit public interest research and advocacy organisation based in New Delhi, assessed the performance of 20 cities for waste segregation at source, transportation, waste processing, and adoption of decentralised systems. In this assessment, these two cities won the four-leaf award and were ranked second in position, after Vengurla, which is a town in the Sindhudurg district of Maharashtra. Under the "Clean Home, Clean City" (Nirmal Bhavan, Nirmal

Gram) initiative, the Alappuzha established a pipe composting system, an aerobic composting system, and biogas plants. The municipality set up fixed as well as portable biogas plants in both residential and public places. According to the renowned magazine *Down to Earth*, this pilot project soon was successful and benefited almost 40,000 households in 52 wards (Agarwal 2017). This progress of waste management was the result of a protest in 2012 in response to the shutting down of the dumping site at Sarvodayapuram, where garbage had been dumped for decades. This protest turned into a movement, and cameras were installed to catch people who litter. Alappuzha saved approximately 11 Lacs Rupees on transportation and liquefied petroleum gas bills. The city was recognised by United Nations Environmental Programme (UNEP), alongside four other anti-pollution cities across the world. The other four cities in the list were Osaka (Japan), Ljubljana (Slovenia), Penang (Malaysia), and Cajica (Colombia) (Agarwal 2017).

In Thiruvananthapuram, composting is being practised at the household level. Thiruvananthapuram consists of 100 wards, and each ward has 2500–3000 households (Bhardwaj 2017). The residents receive assistance on the technical element of composting. For composting at the source, local service providers put forward a three-layered bin and 30 l coco peat-based inoculum. The local service provider collects only the biodegradable waste. The technician will assist in the operations and also collects the compost if it is not required by the household and charges 200 Rupees monthly for providing the services (Bhardwaj 2017). The residents are also encouraged to use biodegradable products and to boycott non-biodegradable products through a programme called "Green Protocol", which was launched by the state government of Kerala during the Onam celebrations, which is Kerala's grand festival.

4.5   The Overall Scenario of Composting in India

Organic matter makes up more than 65% of MSW generated in India (NEERI Report 2012). If this organic fraction of MSW is processed by employing technologies like composting and anaerobic digestion, it can result in a reduction of 50% of the load on landfill sites. In India, only 10–12% of the total MSW is processed by means of composting, and the resulting compost possesses a poor nutrient value and mostly contains heavy metals like zinc, copper, lead, cadmium, lead, nickel, and chromium in higher concentrations as compared to the prescribed standard for compost (Hoornweg et al. 2000 and Waste to Energy Research Council 2012). This poor quality of compost is mainly due to the mixed nature of the waste (Hoornweg et al. 2000). Due to a lack of a continuous supply of feedstock to the compost plant and improper management, composting of biodegradable waste is not successful at large scale in India.

Similarly, improper segregation of waste, higher costs involved in the operation and maintenance of composting plants, and higher costs of manure than chemical fertiliser are major reasons for its non-viability in the Indian context (UNEP 2004).

From a field survey, it was observed that many places in India face a failure of composting plants. For example, a windrow composting plant at Durgapur (300 TPD) and Kamarhati (200 TPD), West Bengal, was not able to achieve sales breakeven point due to issues with the waste characteristics; the plants are about to close. In the same state of West Bengal, the vermicompost plant at Chandannagar and Kalyani is performing well, but the vermicompost plants at Panihati (100 TPD), Bhadreshwar (20 TPD), and Khardah (30 TPD) are in poor shape because of the large variation in the waste characteristics and lack of supervision in the quality control of the product. The same situation applies to the vermicomposting plant (50 TPD) at Bidhannagar. The Kolkata Municipal Corporation is operating a 200 tonnes capacity windrow composting plant sporadically, at less than one-fourth of its capacity, due to the variation in waste characteristics. The compost plants at Mangalore (300 TPD) and Mysore (400 TPD) in the state of Karnataka and a 500 TPD compost plant situated at Nasik in the state of Maharashtra are also processing at lower plant capacity (Aich and Ghosh 2019).

5   Capacity Building Efforts: Strategies and Schemes Launched by the Government of India

Indian MSW management has good potential for composting as waste because it contains a significant amount of organic waste. A decentralised composting system is an appropriate option for India to efficiently run a composting unit. The Indian Government is also encouraging decentralised composting systems, since many environmentalists, policymakers, and researchers have advised that composting is the most feasible and effective waste management technique, in comparison to other techniques (Aich and Ghosh 2019). The government has also developed schemes for MSW capacity building in the country (Ministry of Chemicals and Fertilisers (MOC&F) 2017). Door-to-door waste collection and segregation at the source aid in the collection of waste that has similar characteristics and its disposal at the right place in an economical and efficient manner. A Market Development Assistance scheme has also been launched to enhance capacity building for the compost market. The Indian Government is encouraging all stages of capacity building for city compost. In the last few years, all countries throughout the globe have begun developing strategies for the management of solid waste in a sustainable manner. Composting, as a part of integrated solid waste management, is effective in almost all scenarios. In view of this, the Government of India has also framed different policies and strategies to encourage the practice of composting in India. Schemes like "Swachh Bharat Mission" promote the concept of cleaner and greener cities. Most of the challenges related to the efficient management of MSW in the country are stated in the CPHEEO (2016). The management of solid waste in the country is among the key factors in urban development, and for this reason, the Government of India has launched different schemes, including the Atal Mission for Rejuvenation and Urban Transformation (AMRUT), Jawaharlal Nehru National Urban Renewal Mission (JNNURM), Swachh Bharat Mission, and Smart Cities Mission. In these schemes, solid waste management is generally associated with the water supply system, storm drain system, and sewerage system. Regarding organic waste management, in particular, one recent report, "34th Report on Implementation of Policy on Promotion of City Compost", was presented in the Parliament (Lok Sabha) on 10 April 2017. This report addressed one of the most serious problems associated with urbanisation. According to this report, the stated framework of City Compost is explained in the following section:

- The "Swachh Bharat Mission" scheme is very closely associated with organic waste processing and its use as compost. Hence it can be considered as a rational part of this scheme.
- The composting process helps in reducing the amount of waste at landfill sites.
- Composting also prevents GHG emissions and harmful toxic materials from leaching into groundwater.
- The organic carbon constituent of compost is useful in maintaining and improving soil fertility.
- According to the Ministry of Urban Development, the entire potential of city compost plant in India is 0.71 MT per year, while the current production is 0.15 MT per year.

In the view of this, the MOC&F of the Government of India released a notification in Parliament (Lok Sabha), dated 10 February 2016, to promote city compost schemes (MOC&F 2017). The key features of the scheme are as follows:

(i) An amount of 1500/– Rupees will be provided per tonnes of city compost in the form of marketing development, which in turn will encourage the manufacturing and marketing of the city compost.

(ii) At the initial stage, the existing fertiliser manufacturer will be promoted by the marketing of city compost. Meanwhile, other marketing entities marked by concerned state governments may also be involved, with the approval of the Department of Fertilisers.

(iii) The manufacturer involved in the promotion and sale of city compost may get a subsidy as per the decision of the Department of Fertilisers.

(iv) A fertiliser manufacturer can co-market city compost with chemical fertilisers through their network in the market.

(v) Companies will adopt the village to encourage city compost utilisation.

(vi) The government and public sectors in India will also promote the use of city compost for their gardens and horticulture, to the greatest possible extent.

(vii) Farmers should be educated about the benefits of city compost through information, education, and communication campaign conducted by the Department of Agriculture, Cooperation, and Farmers' Welfare. Agricultural universities will also participate in the same.

(viii) The Ministry of Urban Development will look after installations of the compost plants across the state.

(ix) A Bureau of Indian Standards (BIS) or eco-mark will be developed with the association of BIS to ensure good quality of compost and so its acceptance among farmers.

(x) Government bodies like the Department of Agriculture and Fertilisers and the Ministry of Urban Development will be involved in monitoring to provide information on the availability of city compost. This mechanism will help compost manufacturers and fertiliser marketing concerning coordination issues.

(xi) Expenditure on market development assistance will be met from the budget provision for the Department of Fertilisers.

As a result of this notification, fertiliser companies have adopted 100 villages to promote the city compost. After the implementation of this policy, the production of compost plants is expected to increase, and the target is the processing of 100% of MSW. Significant results are expected to be seen in a year (MOC&F 2017). Decentralised composting units are a feasible option, but skilled manpower is required in good strength. The Indian Government is helping right from waste pickers to farmers with appropriate schemes for the efficient management of organic waste in a sustainable manner.

The Indian Government is also taking steps towards spreading awareness among citizens and by providing e-learning for MSW practices. These courses involve case studies to explain the appropriate options for particular situations. Specialised courses on composting are included in the subcategory of the 200 series (205, 206, 214, 217, 230, 235, and 252) (CPHEEO 2016). This course has been designed to systematically enhance capacity building on composting in India. These courses comprise several case studies of Indian cities, such as Vijayawada, Coimbatore, Bengaluru, Thiruvananthapuram, and Kochi. Apart from the case studies, these courses offer training on choosing suitable composting facilities (CPHEEO 2016).

6 Conclusion and Recommendations

Composting has been practised in India for a long time. However, the extent of its use is surging with time and need. At present, waste management has become a major challenge due to the unscientific practices of waste management observed in the past few decades. Composting technology can be successful or may fail: this depends on its planning and execution. Decentralised composting is burgeoning in India, as the management of smaller units is comparatively easier than bigger units, and the installation cost also affects the financial sustainability of the operational plant. Implementing composting plants of greater capacity generally requires huge investments, and recovering the invested amounts becomes difficult; thus, the losses incurred may lead to the closure of the plants. As discussed earlier, the installation of decentralised smaller composting units is effective at the ground level (MHUA 2018).

One of the major factors in the sustenance of composting plants is community participation. The active participation of the general masses is extremely important; if waste is segregated at the source itself, then the management of waste can be done most effectively. Earlier, the Indian population was negligent towards waste segregation; however, changes in behaviour started when Indians started experiencing the unpleasant views, odours, and health issues that arise due to growing volumes of MSW left unmanaged. Moreover, awareness of the health risks related to unprofessional MSW management has increased. Cities like Alappuzha, Thiruvananthapuram, Mysuru, Panjim, and Panchgani are progressing towards the concept of zero waste discharge and becoming self-sustainable cities. In mega metro cities, adopting composting technology for effective waste management is still a major issue, due to behemoth volumes of waste generation, and in big cities, composting is still a big issue as the amount of waste generated is too huge to manage.

Busy lifestyles, coupled with higher living standards, in the urban parts of the country have also led to negligence regarding the management of waste. However, the critical steps taken by the government to manage generated wastes and put to better use have also resulted in a change in the perspective of the general population. It has been observed that the population is becoming aware of problems associated with effective waste management and is becoming more conscious regarding the effectual management of waste. People are sensitised about this topic, and they are trying to involve themselves more in waste management activities.

Adopting composting techniques for processing organic waste has shown a fair reduction in emissions, as well as in the cost involved in different waste management techniques. Cities with successful operation composting facilities are inspiring other cities to adopt the same, thus resulting in a reduction in their investment and emissions. Apart from these benefits, the general population is also experiencing health benefits, which are also extremely important. If the metro cities plan to install large-scale composting units, precise planning for the recovery of cost should be done in advance by forming various market strategies. It will always prove beneficial to install and operate a pilot-scale composting unit rather than directly going for large-scale composting plants. A reconnaissance study of the concerned region, along with the characteristics of the waste and the acceptance of the technology, should be carried out before the installation of the composting plants. Different urban parts of the country are gradually coming forward to join the initiative for sustainable waste management in India. The government is also encouraging them to manage their waste by providing technical assistance, subsidies for plant installation, and a compost market. The government schemes and strategies are also playing a paramount role in successful compost plant operation.

As described in previous sections, several schemes supporting the composting of organic waste generated from the urban regions of the country have been launched and promoted by the Government of India. Despite all the policies and schemes framed by the government, their implementation on the ground level is still a major challenge. From experience, it has been found that the ruling government has tried to implicate all the perspectives of a composting technique for making it feasible. Different states in the country have different perspectives on and different visions of composting. Sates which have been practising composting for a few years have a commercial approach, while states in the country which are planning to adopt composting to manage their waste have a sustainable approach, ultimately resulting in sustainable development. The efforts taken by the government to promote composting technology should be appreciated, and the policies should be strictly followed to ensure their successful implementation. It is the fundamental duty of every citizen to consider their country as their own home and to take care to keep it *green and clean.*

References

Adak, B. (2018). North Delhi to get green waste compost plant. *India Today.* Available at: https://www.indiatoday.in/mail-today/story/north-delhi-to-get-green-waste-compost-plant-1283362-2018-07-12. Accessed 3 Oct 2018.

Agarwal, R. (2017). Alappuzha gets recognised by UNEP for its solid waste management practices. *DowntoEarth.* Available at: https://www.downtoearth.org.in/news/alappuzha-gets-recognized-by-unep-for-its-solid-waste-management-practices-59258. Accessed 3 Oct 2018.

Aich, A., & Ghosh, S. K. (2019). Conceptual framework for municipal solid waste processing and disposal system in India. In *Waste management and resource efficiency* (pp. 91–107). Singapore: Springer.

Basu, T. (2014). New law of the land: Compost or perish, *The Hindu Business Line.* Available at: https://www.thehindubusinessline.com/news/variety/new-law-of-the-land-compost-or-perish/article23030221.ece. Accessed 2 Oct 2018.

Belyaeva, O. N., & Haynes, R. J. (2009). Chemical, microbial and physical properties of manufactured soils produced by co-composting municipal green waste with coal fly ash. *Bioresource Technology, 100,* 5203–5209.

Bhardwaj, D. (2017). With the help of residents, this is how Trivandrum is composting its own waste. *The Better India.*

Bhattacharya, S. (2016). A tiny town in West Bengal is turning waste into piles of wealth. *Hindustan Times.*

Census. (2011). *Primary census abstracts.* Registrar General of India, Ministry of Home Affairs, Government of India.

CPCB. (2013a). CPCB annual report 2011–12. *Ministry of environment.* Forest and Climate Change, Government of India.

CPCB. (2013b). *Status report on municipal solid waste management 1–13.* Ministry of Environment, Forest and Climate Change, Government of India.

CPHEEO. (2016). *Swachh Bharat mission municipal solid waste management manual.* Central Public Health & Environmental Engineering Organisation. Ministry of Housing and Urban Affairs, Government of India.

Desai, K. (2018). How these temples give a new life to old flowers. *The Times of India.* Available at: https://timesofindia.indiatimes.com/india/how-these-temples-give-a-new-life-to-old-flowers/articleshow/65762167.cms. Accessed 3 Oct 2018.

Diener, S., Zurbrügg, C., & Tockner, K. (2015). Bioaccumulation of heavy metals in the black soldier fly, Hermetia illucens and effects on its life cycle. *Journal of Insects as Food and Feed, 1,* 261–270.

Dinesh, M. S., Geetha, K. S., Venugopalan, V., Kale, R. D., & Murthy, V. K. (2010). Ecofriendly treatment of biomedical wastes using epigeic earthworms. *Journal of Indian Society of Hospital Waste Management, 9*(1).

Dutta, S., (2017). *From composting garden waste to treating 200 tonnes of waste daily, Nagpur's techniques show its all round efforts in improving waste management.* Swachh India. Accessed 3 Oct 2018.

Dutta, S. (2018). *Kolkata municipal corporation's waste management practices come under state pollution control board's scanner.* Swachh India. Accessed 2 Oct 2018.

Fang, M., Wong, J. W. C., Ma, K. K., & Wong, M. H. (1999). Composting of sewage sludge and coal fly ash: Nutrient transformations. *Bioresource Technology, 67*, 19–24.

Gajalakshmi, S., & Abbasi, S. A. (2008). Solid waste management by composting: State of the art. *Critical Reviews in Environmental Science and Technology., 38*(5), 311–400.

Halder, R. (2017). Five colonies of Central Delhi to turn zero-waste by 2018. *Hindustan Times.*

Hoornweg, D., & Bhada-Tata, P. (2012). *What a waste: A global review of solid waste management. ANNEX J: MSW generation by country — Current data and projections for 2025* (Urban Development Series). Washington, DC: The World Bank.

Hoornweg, D., Thomas, L., & Otten, L. (2000). *Composting and its applicability in developing countries* (Working Paper Series 8). Washington DC: The World Bank.

Hubbe, M. A., Nazhad, M., & Sánchez, C. (2010). Composting as a way to convert cellulosic biomass and organic waste into high-value soil amendments: A review. *BioResources, 5*, 2808–2854.

Jouraiphy, A., Amir, S., El Gharous, M., Revel, J. C., & Hafidi, M. (2005). Chemical and spectroscopic analysis of organic matter transformation during composting of sewage sludge and green plant waste. *International Biodeterioration and Biodegradation, 56*, 101–108.

Karthigeyan, M., & Alagesan, P. (2011). Millipede composting: A novel method for organic waste recycling. *Recent Research in Science and Technology, 3*(9), 62–67.

Kumar, S. (2011). Composting of municipal solid waste. *Critical Reviews in Biotechnology, 31*, 112–136.

Kumar, S., Bhattacharyya, J. K., Vaidya, A. N., Chakrabarti, T., Devotta, S., & Akolkar, A. B. (2009). Assessment of the status of municipal solid waste management in metro cities, state capitals, class I cities, and class II towns in India: An insight. *Waste Management, 29*, 883–895.

Kumar, S., Negi, S., Mandpe, A., Singh, R. V., & Hussain, A. (2018). Rapid composting techniques in Indian context and utilization of black soldier fly for enhanced decomposition of biodegradable wastes: A comprehensive review. *Journal of Environmental Management, 227*, 189–199.

Lou, X. F., & Nair, J. (2009). The impact of landfilling and composting on greenhouse gas emissions: A review. *Bioresource Technology, 100*, 3792–3798.

Manyapu, V., Mandpe, A., & Kumar, S. (2017). Synergistic effect of fly ash in in-vessel composting of biomass and kitchen waste. *Bioresource Technology, 251*, 114–120.

Ministry of Housing and Urban Affairs (MHUA). (2018). *Urban advisory on on-site and decentralized composting of municipal organic waste.* Swachh Bharat Mission.

MOC&F. (2017). *Standing committee on chemicals and fertilizers (2014–15).*

Monson, C. C., & Murugappan, A. (2010). Developing optimal combination of bulking agents in an in-vessel composting of vegetable waste. *E-Journal of Chemistry, 7*, 93–100.

Mudgal, S., (2015). Make wealth from waste, *DowntoEarth.* Available at: https://www.downtoearth.org.in/coverage/waste/make-wealth-from-waste-47164. Accessed 3 Oct 2018.

National Environmental Engineering Research Institute (NEERI) Report. (2012). *Optimisation of organic waste to energy systems in India.*

Planning Commission. (2014). Report of the task force on waste to energy: Volume I. *Task Force Waste to Energy, I*, 1–178.

Ramachandra, T. V., Bharath, H. A., Kulkarni, G., & Han, S. S. (2018). Municipal solid waste: Generation, composition and GHG emissions in Bangalore, India. *Renewable and Sustainable Energy Reviews, 82*, 1122–1136.

Rao, P.V. (2016). A small town in West Bengal helped Kolkata win a global award for waste management. *The Better India.*

Sharholy, M., Ahmad, K., Mahmood, G., & Trivedi, R. C. (2008). Municipal solid waste management in Indian cities: A review. *Waste Management, 28*, 459–467.

Times News Network. (2013). Only one-tenth of waste recycled in Kolkata: Study. *The Times of India.*

Times News Network. (2017). Nagpur Municipal Corporation to convert garden waste into compost. *The Times of India.*

Tomić, T., & Schneider, D. R. (2017). Municipal solid waste system analysis through energy consumption and return approach. *Journal of Environmental Management, 203*, 973–987.

UNEP. (2004). State of waste management in South East Asia..

Vandergheynst, J. (2009). *Compost processing and its relationship to compost quality.* Biological and Agricultural Engineering, UC Davis.

Varma, S. V., & Kalamdhad, A. S. (2013). Composting of Municipal Solid Waste (MSW) mixed with cattle manure. *International Journal of Environmental Sciences, 3*(6), 2068–2079.

Waste to Energy Research Council. (2012). *Sustainable solid waste management in India.* Earth Engineering Centre Columbia.

Wei, Y., Li, J., Shi, D., Liu, G., Zhao, Y., & Shimaoka, T. (2017). Environmental challenges impeding the composting of biodegradable municipal solid waste: A critical review. *Resources, Conservation and Recycling, 122*, 51–65.

3

Organic Waste Composting at Versalles: An Alternative that Contributes to the Economic, Social and Environmental Well-Being of Stakeholders

Luis Fernando Marmolejo-Rebellón, Edgar Ricardo Oviedo-Ocaña, and Patricia Torres-Lozada

Abstract Composting is one of the most widely used technologies for the recovery and use of organic waste from municipal solid waste (MSW); however, its implementation in some developing countries has mostly been ineffective. This chapter documents the experience of the composting of municipal organic waste in the urban area of the municipality of Versalles, Valle del Cauca, Colombia. Within the locality, composting of organic waste occurs at an MSW management plant (SWMP), after being separated at the source and selectively collected. The information presented was generated through collaborative research projects, conducted with the cooperation of Camino Verde APC (a community-based organisation providing sanitation services) and Universidad del Valle (Cali, Colombia). The evaluations undertaken show that (i) within the locality, high rates of separation, at the source, in conjunction with selective collection and efficient waste sorting and classification processes in the SWMP, have significantly facilitated the composting process; (ii) the incorporation of locally available amendment or bulking materials (e.g. star grass and cane bagasse) improves the physicochemical quality of the processed organic waste and favours development (i.e. a reduction in process time), leading to an improvement in product quality; (iii) the operation, maintenance and monitoring of the composting process can be carried out by previously trained local human talent; and (iv) revenues from the sale of the final product (compost) are not sufficient to cover the operating costs of the composting process. Despite this current lack of financial viability, the application of technology entails environmental benefits (e.g. a reduction in the generation of greenhouse gases) and social benefits (e.g.

L. F. Marmolejo-Rebellón (✉) · P. Torres-Lozada
Faculty of Engineering, Universidad del Valle, Cali, Colombia
e-mail: luis.marmolejo@correounivalle.edu.co; patricia.torres@correounivalle.edu.co

E. R. Oviedo-Ocaña
School of Civil Engineering, Universidad Industrial de Santander, Bucaramanga, Colombia
e-mail: eroviedo@uis.edu.co

employment opportunities), which, given the conditions in the municipality studied, highlight the relevance of this technological option.

Keywords Amendment materials · Bulking materials · Selective collection · Source separation · Zero waste

1 Introduction

Composting has become one of the most popular (and frequently utilised) technologies for the recovery and use of organic waste, which is the predominant fraction of municipal solid waste (MSW) in smaller municipalities in developing countries. The main advantages over other technological options are its low cost and operational simplicity (Li et al. 2013, Sundberg and Navia 2014). In addition, compost contains nutrients which can help improve levels of soil organic matter and soil fertility status (Hargreaves et al. 2008). However, experiences in developing countries (Barreira et al. 2006, Ekelund and Nyström 2007, Organización Panamericana de la Salud (OPS) 2005, Zurbrügg et al. 2005), including studies from Colombia (Marmolejo 2011, Superintendencia de Servicios Públicos Domiciliarios 2008), show that the implementation of composting has so far been ineffective. Smaller municipalities, i.e. those with less than 15,000 inhabitants as classified by OPS (2010), present the most critical conditions, which can even lead to the generation of negative environmental and health impacts (e.g. gaseous emissions, odours and soil contamination when the product is applied), ultimately leading to the closure of composting facilities, putting investments at risk.

This chapter details the experience of organic waste management in the minor Colombian municipality of Versalles (Valle del Cauca, Colombia). The focus of the chapter is on the local community's experience, as documented, and the partnership between the company providing waste management services and the research group 'Estudio y Control de la Contaminación Ambiental' (Study and Control of Environmental Pollution) at Universidad del Valle, Cali.

This chapter includes four sections: Sect. describes the municipality's location (i.e. geographic position, size, climate); Sect. presents the management of the MSW within the locality (i.e. generation and composition of MSW, storage, collection and transport, recovery and recycling and final disposal in a landfill site); Sect. focuses on documenting the composting experience of organic waste in the local-ity, highlighting the technical aspects, technical improvement strategies and the social and economic aspects related to the MSW management; and Sect. presents final considerations arising from the study.

2   General Aspects of the Locality

Versalles is a Colombian municipality, located in the department of Valle del Cauca, at 1860 m above mean sea level, on the western branch of the Andes. Its geographic coordinates are latitude 4°, 34′,43′ and longitude 76°, 12′,23′. The average temperature is 18 °C, with a maximum of 24 °C and a minimum of 12 °C, and it covers an approximate area of 352 km², of which 246 km² corresponds to the urban area. Versalles is approximately 205 km from Santiago de Cali, the capital of the Valle del Cauca department and Colombia's third largest city in terms of population. The municipality has an urban population of approximately 3831 inhabitants. Figure presents the location of the municipality.

Rooted within its population, Versalles has a broad participatory culture, as reflected in events and initiatives promoting community consolidation. An example of this was the opening of an access road at the beginning of the twentieth century, which resulted from so-called 'minga' activities – an ancient tradition of collective work for the purpose of social utility. This culture was also promoted by the operation of an educational institution which offered a modality of social promotion, in which many of the social leaders, who have been promoting local development in the last 25 years, were trained.

Additionally, in the 1980s, under the leadership and guidance of initiatives such as the Committee of Community Participation in Health, the municipality advanced

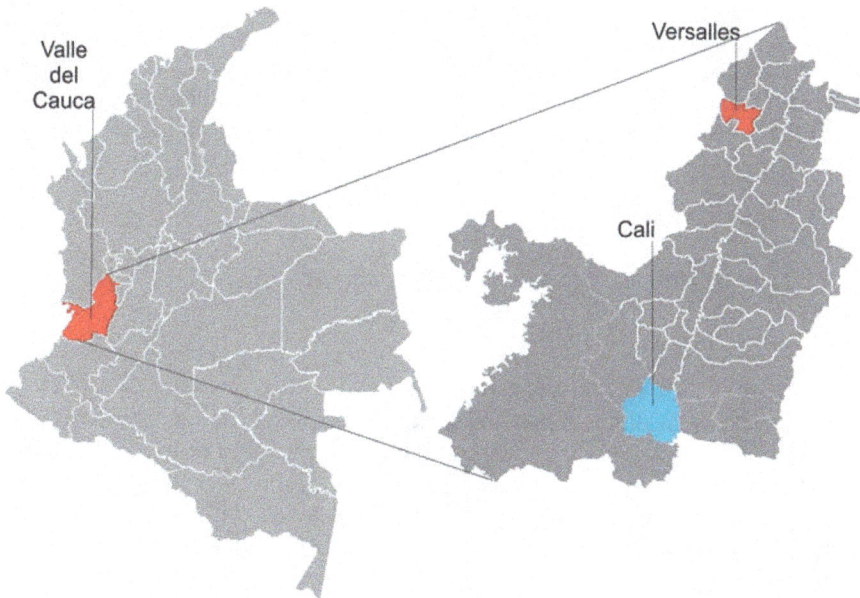

Fig.   Location of the municipality of Versalles, Valle del Cauca, Colombia

planning processes aimed at addressing urgent needs in the fields of health, education and culture, agricultural development and the environment, infrastructure and public services, business management, security and peaceful coexistence. Improvements made in these areas have earned Versalles recognition, such as the distinction of 'Healthy Municipality for Peace', awarded by the Pan American Health Organization in 1993 (Restrepo 2002).

Another example of a social scheme of local empowerment is the provision of public water, sewerage and sanitation services, provided by the Administracion Pública Cooperativa (APC) Camino Verde Company, an organisation whose board of directors consists of 51% of members from community-based organisations and 49% from the mayor's office. The internal organisation of the company with respect to both administrative and operational functions is clearly differentiated and inter-related. For example, its personnel plant is made up mainly of people from the locality, who actively participate in continuous improvement programmes promoted by the company, thereby leading to high levels of company service and user satisfaction.

3   Management of MSW

The waste management service in Versalles has a significant connotation for the country, because the management scheme is oriented towards both a reduction in MSW generation and the subsequent recovery and recycling of any MSW generated. This scheme results in the recovery of approximately 80% of all the MSW generated and, in comparison to populations with similar conditions, leads to a lower per capita generation of MSW (Marmolejo 2011). The following is a description of the waste management scheme in the locality, focusing on the recovery and use of organic waste, which is the subject of this chapter.

3.1   Generation and Composition of the MSW

In sampling and characterisation studies conducted by Erazo and Pereira (2010) and Vélez (2016), the production per capita (ppc) of MSW was estimated to be 0.29 kg/inhabitant-day in the urban area of the town. In the first study, it was established that the ppc in the residential sector was 0.27 kg/inhabitant-day, a value similar to that found in later studies of the same locality, also focused on the residential sector (0.29 and 0.24 kg/inhabitant-day), by Pérez and Reina (2015) and Maldonado (2015), respectively. These ppc values are lower than the suggested minimum value for Colombian populations (0.30 kg/inhabitant-day) with conditions similar to those of Versalles, as documented in Title F of the 'Reglamento Técnico del Sector Agua

Table Production of MSW in Versalles (Valle del Cauca)	Material	Proportion (%)	Quantity (t/ month)
	Organic waste	65.8	28.6
	Recyclable materials	17.2	7.5
	Non-usable waste	17.0	7.4

Potable y Saneamiento Básico' (Technical Regulations for the Drinking Water and Basic Sanitation Sector) – RAS (Ministerio de Vivienda Ciudad y Territorio 2012).

Research carried out seeking to explain this situation found that the majority of local waste management service users had a zero waste-oriented culture, associated with (i) rural cultural practices, such as the re-use of food waste (e.g. for animal feed, or as soil fertiliser); (ii) the existence of other strategies for the recovery and in situ valorisation of recyclable materials, promoted by Camino Verde APC and by educational institutions; and (iii) the establishment of 'responsible consumption' practices by some of the residents.

In relation to the physical composition of MSW, organic waste predominates, accounting for 65.8%, followed by recyclable materials, accounting for 17.2% (i.e. paper, cardboard, plastic, metals and glass), and finally, non-usable or rejects (i.e. hygienic material, sweepings and others not recyclable) that reach 17.0% of the MSW. This composition is similar to that reported by Marmolejo et al. (2009) for other municipalities near to the town studied and is consistent with the trend found in other developing countries (Aalok et al. 2008; Thi et al. 2015).

3.2 Storage of MSW

Source separation of waste is a common practice in the locality, applied by a high proportion of users (i.e. varying between 60% and 90% of users). This variation is associated with factors such as (i) the discontinuity of the awareness activities developed by Camino Verde APC, which on occasion have been reduced due to budget cuts, and (ii) the hosting of large-scale community events, such as festivals and municipal parties, during which there is an influx of visitors from outside, who are unaware of the social and environmental process commitments existing in the town.

In an evaluation conducted in 2015, Cardona and Pinto (2015) found that 0.9% of residential users stored their waste in only one container, 18.5% in two and the remainder in three or more. Plastic bags, and a combination of plastic bags and containers, are the most used storage options. It is also common to find waste stored in roofed places which limit the conditions required for decomposition and therefore do not affect the quality of the stored materials. Waste is generally put out by the user at the time of the garbage truck collection, or 10 min before.

3.3 Collection and Transport

Waste is regularly collected on the curb of homes or establishments, twice a week, on Mondays and Thursdays. For each solid waste collection day, two trips are usually made to a solid waste management plant (SWMP). A dump truck is used for the collection, which has two uncovered compartments of the same size (organic waste is collected in one compartment and recyclable waste and rejects in the other). The collection is carried out by four operators, two of whom advance through the streets, whilst the other two receive and accommodate the waste in the respective compartments of the dump truck.

3.4 Recovery and Recycling

Once the collection route is concluded, the vehicle goes to the SWMP, where activities are carried out for the recovery and recycling of the MSW. The SWMP is located 1 km from the urban limits. The SWMP has an area of 1.5 hectares (ha), with slope inclination ranging between 5% and 50% (Oviedo 2015). The operating scheme of the SWMP is shown in Fig..

The waste management in the SWMP is conducted in the following steps:

(a) Reception of materials: Includes the weighing of the loaded and unloaded collection vehicle and the emptying of the solid waste in the areas destined for recovery or temporary storage.

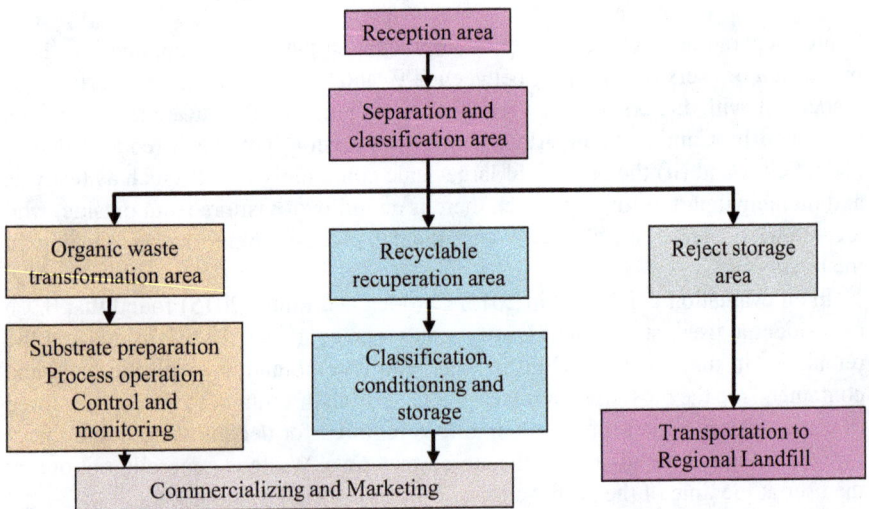

Fig. Scheme of operation of the SWMP

(b) Classification and packaging of recyclable materials: Here the separation of plastic, paper, cardboard, glass and metals takes place. Paper, cardboard and plastic bags are compacted.

(c) Transformation of organic waste: Initially, when the plant was first opened, organic waste recovery and valorisation was carried out using vermicomposting (Corporación Autónoma Regional del Valle del Cauca 2005). However, in 2008, after an evaluation that included the monitoring of product quality, Camino Verde APC opted to change to manual composting. Unfortunately, the difficulties associated with the manual turning and monitoring of the compost have limited its application.

(d) Temporary storage of waste to be sent to final disposal (rejects): Waste that cannot be used for various reasons, such as its objectionable quality or lack of demand in the regional market, is temporarily stored in the plant and then sent to a landfill site located approximately 90 km from the town. The transportation is carried out once every 2 weeks in the same vehicle as is used for collection.

Figure presents the waste stream in the locality. It can be observed that organic waste is the largest fraction that enters the SWMP (65.8%). Recyclable materials, despite representing 17.2% of the generated MSW, enter the SWMP in a smaller amount (5.0%), due to the fact that 12.2% (1.26 t/week) is being taken by local waste pickers. Of the total MSW received in the SWMP, only 17.5% is transported to the final disposal site. These yields are associated with the quality of the raw materials of the different processes, which come from source separation

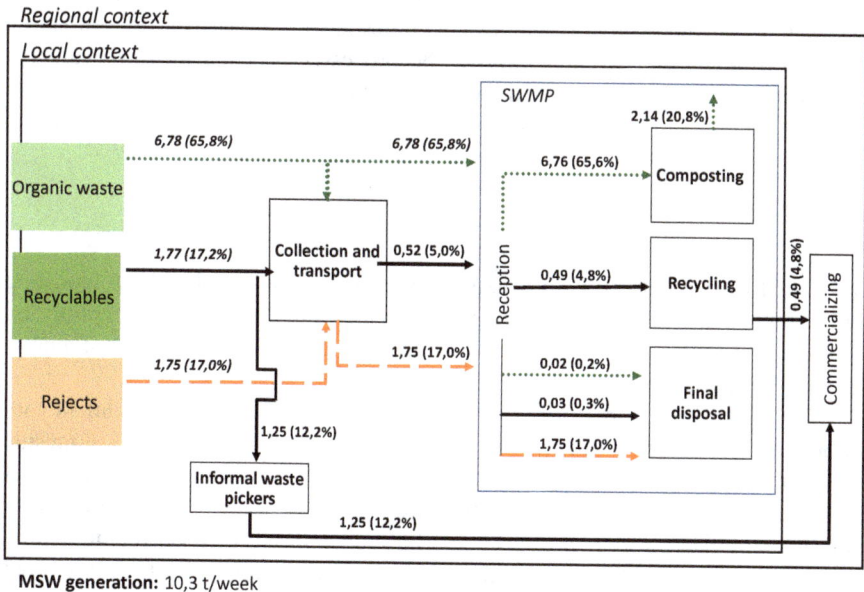

MSW generation: 10,3 t/week

Fig. MSW flow in Versalles. (Source: Adapted from Erazo and Pereira 2010)

practices and selective collection of MSW, as well as efficient waste sorting and classification processes in the SWMP.

4   Composting of Organic Waste

In 2008, Camino Verde APC and Universidad del Valle established cooperation links that had the purpose of contributing to the improvement of the waste management service in the population. The evaluation of SWMP functioning was included amongst the jointly agreed priorities, and within this priority, the evaluation of vermicomposting (the organic waste transformation process applied at that time) was included, in order to comply with the regulations and standards in Colombia and to obtain the necessary approvals to market the compost product.

The evaluations conducted showed that the vermicomposting, as carried out, did not meet the necessary operation and maintenance requirements. This was reflected in the product quality analysis: the physiochemical and microbiological characteristics did not satisfy the requirements of the Colombian regulations associated with this type of material (i.e. NTC 5167) (Instituto Colombiano de Normas Técnicas y Certificación (ICONTEC) 2004, Marmolejo 2011). Additionally, the technical evaluation showed that, given the waste generation context and conditions at that time, the area of land available in the SWMP for the processing of organic waste was insufficient for vermicomposting. For these reasons, the possibility of utilising composting was evaluated and validated, and research projects were conducted jointly by Camino Verde APC and Universidad del Valle. Within the framework of these projects, technical, economic and social studies have been carried out, evaluating the improvement and sustainability of the process. The main aspects evaluated and the respective results obtained are discussed in Sect..

4.1   Technical Aspects

Taking into account the fact that organic waste is the raw material of the composting process (substrate) and that the raw material quality is highly variable, sampling and characterisation activities were carried out to evaluate both the variability of quality and its effect on the process.

The results showed that there was a need to use co-substrates (amendment and/or bulking materials), in order to improve both the quality of the substrate, alongside the operation and maintenance conditions, and the quality of the product. Given that the completed research sought to contribute to the sustainability of the system, it was essential that the co-substrates could be sourced from the local region, within easy reach of Camino Verde APC, in order to improve the product quality without risking the economic sustainability of the technology.

Table Physicochemical characteristics of organic waste composting substrates

Parameter	Versalles[a]	Recommended values for the start of the process
pH	5.5 ± 0.5	6.5–7.5
Moisture	76.7% ± 3.2%[a]	60
Ashes	25.1% ± 5.6%[b]	–
TOC	33.0% ± 4.8%[b]	–
Total N	1.6% ± 0.5%[b]	>1
C/N	21.7 ± 5.3	25–30
Total P	0.3% ± 0.1%[b]	>1
Total K	1.6% ± 0.5%[b]	>1

Note: [a]Oviedo et al. (2015), obtained from monitoring 39 sessions in the locality under study, [b]on wet matter basis, [c]on dry matter basis

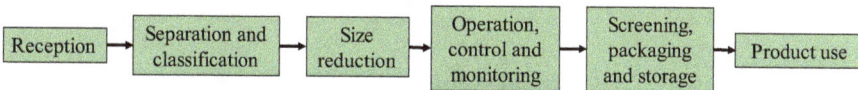

Fig. The compost management process of organic waste in the SWMP

Substrate Conditions

The local organic waste is characterised by a high content of food waste, with 92.9% (±4.3%) being unprocessed foods, amongst which bananas, tubers (32.6%) and citrus fruits (15.5%) are in greater proportion. This waste is characterised by having rapidly degradable organic matter that makes it complex to manage. Regarding the physicochemical composition of organic waste, it can be observed that it presents critical conditions for composting, such as acid pH, excess moisture, deficiencies in total organic carbon (TOC) and total phosphorus (total P) and a lower carbon/nitrogen (C/N) ratio than that recommended at the start of composting.

Process Conditions

Figure presents the manual composting process stages carried out in the SWMP. The waste is discharged in the organic waste area. The operators sort through the waste and extract any impurities or rejects. Then, using tools such as machetes and shovels, the size of some thick materials is reduced. Subsequently, compost piles are formed conically from the waste that enters from each collection day (i.e. Monday and Thursday).

The composting piles are operated (i.e. by manual turning and wetting) and, depending on the operator's time span availability, variables such as temperature, pH and moisture are controlled. The process takes approximately 120 days (i.e. about 90 days in active phase and 30 days in maturation). To establish when the process is complete, the operators perform visual and odour tests on the material; no

Photo Area of the
composting process

Photo Operational
turning activities

other tests are carried out to determine the level of stability or maturity of the processed material. Finally, the material received from the maturation area of the
com-posting process is manually sifted, packed and stored until use. Photos show
details of the manual composting process carried out in the study area.

Product Quality

The evaluation of product quality, as carried out in organic waste composting experiments on a pilot scale in developing countries such as Nigeria, Brazil and India
(Adekunle et al. 2010; Saha et al. 2010), has shown that the composting of municipal organic waste without the use of co-substrates is characterised by high pH and
ash values and limited content of TOC and nutrients such as total P, which
ulti-mately restricts product commercialisation. Table presents the results of
two

Table Evaluations of the product quality in the SWMP

Parameters	Units	Product A[a]	Product B[a]	Norm NTC 5167[b]
pH	Units	10.10 ± 0.28	8.01 ± 0.13	>4 and <9
Moisture	% (w/w)	32.77 ± 5.6	39.00 ± 0.9	<35
TOC	% (w/w)	13.73 ± 0.4	12.77 ± 1.97	>15
Total N	% (w/w)	0.90 ± 0.12	1.54 ± 0.51	>1
C/N	–	14.77 ± 1.9	9.43 ± 5.4	–
Ashes	% (w/w)	65.67 ± 2.3	62.63 ± 1.9	<60
Total K	% (w/w)	3.23 ± 0.38	3.78 ± 0.37	>1
Total P	% (w/w)	1.32 ± 0.06	1.45 ± 0.17	>1
Density	g/cm^3	0.55 ± 0.1	0.61 ± 0.03	<0.6
Water holding capacity	%	120.4 ± 6.9	124.4 ± 7.1	>100
Cationic exchange capacity	meq/100 g	49.7 ± 1.6	52.3 ± 0.9	>30
Faecal coliforms	MPN/g[c]	0.0	17.0	–

Note: [a]Average values of three replicates, [b]Colombian Standard Technique referring to organic products used as soil conditioners, [c]MPN: most probable number

typical product quality evaluations at the SWMP and also compares the results with the values established for the use of the product as an organic amendment in the Colombian Technical Standard, NTC 5167 (ICONTEC 2004).

Improvement Strategies

Given the natural limitations of organic waste for effective composting and the necessity to obtain a final product which satisfies both the regulatory requirements and that of the user within the locality, substrate conditioning options were evaluated. The studies of Marmolejo (2011) and Oviedo et al. (2015) identified potential amendment or bulking materials, including (i) fruit harvest residues, (ii) cane bagasse (CB), (iii) star grass (SG), (iv) material in process (MP), (v) wood ashes (WA), (vi) bovine manure, (vii) pig manure and (viii) chicken faeces.

The selection criteria included the consideration of variables such as quality and quantity of required material, acquisition costs, access and distance to the site of material generation, provision for material delivery and material handling facilities. The amendment or bulking materials chosen were (i) WA, (ii) MP, (iii) BC and (iv) SG, which were evaluated in experiments on a pilot scale of organic waste composting. The evaluation processes are described as follows:

• Addition of WA (proportion of 2%, 4% and 8%): It was favourable to increase the initial pH, improve the nutrient content of the substrates and dampen the acids generated in the first phase of the process. However, it had no effect on accelerating the start of the process, and an excessive increase in pH was observed with the addition of 4% and 8% WA, which, associated with thermophilic temperatures, could cause greater total N losses. The germination tests in the products with WA showed the presence of phytotoxic agents (e.g. presence of salts) that compromised the use of the product (Oviedo et al. 2014a).

- Incorporation of processing material (i.e. material used in the composting process and taken from the maturation stage of the piles) (25% proportion) contributes to the improvement of initial moisture and TOC of the substrates, allowing them to attain higher temperatures in the thermophilic range than composting of only organic waste. However, similar duration times of the thermophilic and cooling stages and the obtaining of a product with similar physicochemical characteristics showed significant differences only for the parameters of C/N and ashes ($P < 0.05$) (Oviedo et al. 2015).

- Addition of SG (34% ratio) and CB (22% ratio), each separately: In comparison with piles of organic waste alone, in both cases the start of the process was accelerated, the duration of the thermophilic stage was reduced and suitable conditions were provided for the sanitisation of the material. However, compounds in these materials that are difficult to break down can be the cause of a longer duration of the cooling and maturation phases. The incorporation of SG and CB (each separately) contributed to an improvement in product quality, by producing levels of TOC, cationic exchange capacity, density and water holding capacity. The use of the obtained products can improve soil properties, such as the retention of nutrients, water and increased microbial activity. The fulfilment of a greater number of quality parameters established in the Colombian Technical Standard (Oviedo et al. 2015) can also be observed. Table shows product quality parameters identified in experiments evaluating the incorporation of SG and CB.

On the other hand, to counteract the high moisture content of the organic waste, the effect of the increased turning frequency in the composting process was also evaluated (i.e. six times per week during the first 4 weeks, compared to a pile with a turning frequency of twice per week; at least one flip per week was performed on each type of pile). This study showed a decrease in the processing time of 20% in

Table Product quality in experiments evaluating the incorporation of SG and CB

Parameters	Unit	A[a]	B[b]	Norm NTC 5167
pH	Units	9.90 ± 0.17	7.38 ± 0.07	>4 and <9
Moisture	% (w/w)	34.50 ± 3.3	49.47 ± 11.1	<35
TOC	% (w/w)	18.87 ± 2.7	17.77 ± 1.46	>15
Total N	% (w/w)	2.02 ± 0.50	0.90 ± 0.69	>1
C/N	–	9.67 ± 3.1	28.23 ± 18.7	–
Ashes	% (w/w)	61.03 ± 1.3	57.23 ± 1.0	<60
Total K	% (w/w)	3.92 ± 0.13	3.11 ± 0.40	>1
Total P	% (w/w)	1.26 ± 0.10	1.04 ± 0.08	>1
Density	g/cm^3	0.34 ± 0.0	0.44 ± 0.07	<0.6
Water holding capacity	%	165.6 ± 18.6	168.83 ± 9.9	>100
Cationic exchange capacity	meq/100 g	50.0 ± 2.7	56.2 ± 1.9	>30

Note: [a]A, 34% SG + 66% organic waste; [b]B, 22% CB + 78% organic waste

the mesophilic, thermophilic and cooling phases. However, a decrease in product nutrient content (i.e. nitrogen and phosphorus) was observed. The results also made it possible to demonstrate that control and monitoring schemes, in accordance with local conditions, can be effective in improving the performance of the process. Additionally, the results pose new research challenges, such as the joint evaluation of the effect of the increase in the turning frequency and of the addition of bulking materials (or of amendment) for the composting process and quality (Oviedo et al. 2014b).

4.2   Economic and Social Aspects

The SWMP operation and maintenance is carried out by four workers, who in turn are part of the collection crew (an activity on which they spend approximately 25% of their time). These workers are part of the Camino Verde APC hourly employee crew, and their employment relationship is carried out in strict compliance with the requirements of Colombian labour legislation. This is an aspect that represents an important added value, since the linkage with local participants in the use and valo-risation of the MSW constitutes a source of formal employment in a locality that has few employment options, which is one of its biggest social problems and one of the causes of out-migration of its native population.

The implementation of composting brought with it an increase in the labour requirements, relating to monitoring, operation and maintenance activities, and therefore an increase in costs. With resources obtained through the research projects funded by the Universidad del Valle and Camino Verde APC, the four operators at the plant were trained in aspects related to the composting monitoring and operation process. When reviewing the work assignment of these workers, it was identified that they had time to support the turning and wetting, but not for the monitoring of the process, for which it was necessary to train and link an additional full-time operator (now no longer employed). For a period of time, this last operator was responsible for the monitoring and recording of temperature, pH and pile moisture. Following parameter interpretation and evaluation, the operator was required to make decisions on fundamental aspects of the process such as turning and wetting. The execution of these last activities (i.e. the turning and wetting) was/is the respon-sibility of the entire team of workers.

In general, because of the substrate's characteristics, during the first 5 days in which the piles are made, they have to be turned over between two to three times a week. At the same time, a space was established in the SWMP for a laboratory, in which the operator responsible for monitoring was able to perform moisture and pH determinations using field equipment facilitated by the university and was able to register information about the process. This was successful: the operator responsi-ble for this function took the knowledge into his own hands and made correct deci-sions, as reflected in product quality improvement in terms of variables such as moisture, pH and C/N ratio, a decrease in offensive odours and a reduction in the

Photo Process
monitoring

Photo Final product
storage area

process time. Nevertheless, once the investigation project budget – financed by Universidad del Valle – had been used up, the enterprise was unable to sustain the funding of this operator, and thereafter the activity was suspended. The operators responsible for the maintenance of the plant continued with the turnover and water-ing, insofar as the piles were being moved.

In a cost evaluation of the sanitation service performed by Yusti (2017) it was estimated that the monthly cost associated with the payment of an employee's salary was $1,134,383.00 Colombian pesos (COP) (approximately US$384.40).

The COP monthly cost of endowment was $15,542 (US$5.30) and the COP pro-vision of security elements amounted to $58,667.00 (US$19.90).

The cost associated with the use of supplies was $62,333.00 COP (US$21.10). According to this information, the monthly operation and maintenance costs of the SWMP would sum up to about $5,083,700 COP (US$1723.00).

The final composting product is not currently for sale, since an authorisation for use has not been requested from the competent government agency, and therefore the product is used in internal activities at the SWMP, or for the recovery of eroded soils. In the region, the sale price of a bulk of 40 kg of compost is approximately $8000.00 COP (US$2.71). If the total monthly production of compost (approximately 8360 kg) were to be sold, the income would be approximately $1,672,000.00 COP (US$566.60). With respect to the monthly income from the sale of recovered recyclable materials, Yusti (2017) reported it was in the order of $1,415,525.00 COP (US$479.70). As can be observed, even if all the compost produced was marketed, the SWMP would not be financially sustainable. However, it is important to take into consideration that composting the organic waste at the SWMP reduces the costs of final disposal. Specifically, it avoids four monthly solid waste collection trips, which represent a monthly saving of $696,000.00 COP (US$235.80) in terms of fuel costs and tolls and $728,295.00 COP (US$246.8) in final disposal fee payments.

In 2017, Flórez (2017) carried out a comparison of the environmental impacts generated by the composting of organic waste in the locality, versus the shipment to final disposal at a regional landfill site, using the software SimaPro 8.0 (developed by the company PRé). Overall, Flórez found that in terms of climate change, the impact of final disposal in landfill was approximately 7.5 times that of composting, due to the negative environmental effects of the eutrophication of water (2×) and particulate material formation (5×). In accordance with the above, the use and recovery of waste generated are the most convenient strategies for the municipality of Versalles. Emphasis is placed on this result, specifically in relation to (i) a reduction in generation, (ii) separation at source, (iii) selective collection and (iv) the encouragement to use and enhance MSW.

5   Final Considerations

The integration of key elements of solid waste management, such as community participation and business development in the provision of public services, and the implementation of technologies adapted to the local context have been fundamental pillars of the experience of waste management in the urban area of Versalles.

The predominant fraction of MSW is organic waste, which has a high potential for biological treatment. However, the quality of the organic waste itself presents deficiencies for the composting process, generating products that do not meet quality standards, limiting their use and marketing. Therefore, strategies adapted to the context, such as the incorporation of amendment and bulking materials, can contribute to improving the quality of both the substrate and product, as well as process efficiency. However, it is necessary to continue to strengthen research and development on pilot-scale projects in the context of smaller populations, which can contribute sustainable improvements to the operation of organic waste composting facilities.

In spite of the difficulties of using MSW in the locality in regard to financial feasibility and in particular the valorisation of the use of organic waste, the evaluations carried out highlight the social pertinence of these options: they generate employment opportunities at the local level and reduce environmental impact. The evaluations also highlight the generation of a product that has potential in regard to supplying organic matter and nutrients to the soil in the local context.

Acknowledgements The authors thank Universidad del Valle, for financing the research projects; the Cooperativa de Servicios Públicos de Versalles, Camino Verde APC, for the support provided in the experimental activities; and the undergraduate and MSc students at Universidad del Valle, who participated in the projects.

References

Aalok, A., Tripathi, A., & Soni, P. (2008). Vermicomposting: A better option for organic solid waste management. *Journal of Human Ecology, 24,* 59–64.

Adekunle, I., Adekunle, A., Akintokun, A., Akintokun, P., & Arowolo, T. (2010). Recycling of organic wastes through composting for land applications: A Nigerian experience. *Waste Management & Research, 29,* 582–593.

Barreira, L., Philippi, A., & Rodrigues, M. (2006). Usinas de compostagem do estado de Sao Paulo qualidade dos compostos e processos de producao. *Engenharia Sanitária e Ambiental, 11,* 385–393.

Cardona, Y., & Pinto, A. (2015). *Análisis de la influencia del almacenamiento de los biorresiduos sobre las características del sustrato producido en la cabecera municipal de Versalles – Valle del Cauca,* Trabajo de grado del Programa de Ingeniería Sanitaria y Ambiental, Facultad de Ingeniería, Universidad del Valle, Cali, Colombia.

Corporación Autónoma Regional del Valle del Cauca. (2005). *El Manejo Integral de los Residuos Sólidos,* La Experiencia del Municipio de Versalles.

Ekelund, L., & Nyström, K. (2007). *Composting of municipal waste in South Africa.* Upsala: Upsala Universitet.

Erazo, K., & Pereira, J. (2010). *Caracterización del flujo de residuos sólidos en la cabecera de Versalles – Valle del Cauca,* Trabajo de grado del Programa de Ingeniería Sanitaria, Facultad de Ingeniería, Universidad del Valle, Cali, Colombia.

Flórez, H. (2017). *Análisis de Ciclo de Vida de los biorresiduos municipales generados en Versalles, Valle del Cauca,* Trabajo de grado del Programa de Ingeniería Sanitaria y Ambiental, Facultad de Ingeniería, Universidad del Valle, Cali, Colombia.

Hargreaves, J., Adl, M., & Warman, P. (2008). A review of the use of composted municipal solid waste in agriculture. *Agriculture, Ecosystems & Environment, 123,* 1–14.

Instituto Colombiano de Normas Técnicas y Certificación (ICONTEC). (2004). *Productos para la industria agrícola. Productos orgánicos usados como abonos o fertilizantes y enmiendas de suelo,* Norma Técnica Colombiana.

Li, Z., Lu, H., Ren, L., & He, L. (2013). Experimental and modeling approaches for food waste composting: A review. *Chemosphere, 93,* 1247–1257.

Maldonado, N. (2015). *Posibilidades de reducción en la generación o incremento del aprovechamiento de los residuos sólidos residenciales en la cabecera de Versalles, Valle del Cauca,* Trabajo de grado del Programa de Ingeniería Sanitaria y Ambiental, Facultad de Ingeniería, Universidad del Valle, Cali, Colombia.

Marmolejo, L. (2011). *Marco conceptual para la sostenibilidad de los sistemas de aprovechamiento de residuos sólidos en cabeceras municipales menores a 20.000 habitantes del Valle del Cauca,* Tesis Doctoral, Doctorado en Ingeniería, énfasis Ingeniería Sanitaria y Ambiental, Facultad de Ingeniería, Universidad del Valle, Cali, Colombia.

Marmolejo, L., Torres, P., Oviedo, E., Bedoya, D., Amezquita, C., Klinger, R., Alban, F., & Diaz, L. (2009). Flujo de residuos. Elemento base para la sostenibilidad del aprovechamiento de residuos sólidos municipales. *Ingenieria y Competitividad, 11*(2), 79–93.

Ministerio de Vivienda Ciudad y Territorio. (2012). *Reglamento Técnico de Agua y Saneamiento. Título F. Aseo.* República de Colombia.

Organización Panamericana de la Salud (OPS). (2005). *Informe de la Evaluación Regional de los Servicios de Manejo de Residuos Sólidos en América Latina y el Caribe.* Washington, DC.

Organización Panamericana de la Salud (OPS). (2010). *Informe Regional del Proyecto Evaluación Regional del Manejo de Residuos Sólidos Urbanos en América Latina y el Caribe 2010,* Banco Interamericano de Desarrollo, Asociación Interamericana de Ingeniería Sanitaria y Ambiental.

Oviedo, E. (2015). *Estrategias para la optimización del proceso y la calidad del producto del compostaje de biorresiduos en municipios menores de países en desarrollo,* Tesis Doctoral, Doctorado en Ingeniería, énfasis Ingeniería Sanitaria y Ambiental, Facultad de Ingeniería, Universidad del Valle, Cali, Colombia.

Oviedo, E., Marmolejo, L., & Torres, P. (2014a). Evaluation of wood ashes addition for the pH control of substrates in municipal biowaste composting. *Ingeniería, Investigación y Tecnología, 15*(3), 469–478.

Oviedo, E., Marmolejo, L., & Torres, P. (2014b). Influencia de la frecuencia de volteo para el control de la humedad de los sustratos en el compostaje de biorresiduos de origen municipal. *Revista Internacional de Contaminacion Ambiental, 30*(1), 91–100.

Oviedo, E., Marmolejo, L., Torres, P., Daza, M., Andrade, M., Torres, W., & Abonía, R. (2015). Effect of adding bulking materials over the composting process of municipal solid biowastes. *Chilean Journal of Agricultural Research, 75*(4), 472–480.

Pérez, N., and Reina, M. (2015). *Alternativas para la gestión de residuos peligrosos contenidos en los residuos sólidos residenciales generados en la cabecera de Versalles -Valle del Cauca,* Trabajo de grado del Programa de Ingeniería Sanitaria y Ambiental, Facultad de Ingeniería, Universidad del Valle, Cali, Colombia.

Restrepo, H. (2002). Experiencia del municipio de Versalles, departamento del Valle: una mirada desde la promoción de la salud. *Revista Facultad Nacional de Salud Públic, 20*(1), 135–144.

Saha, J., Panwar, N., & Singh, M. (2010). An assessment of municipal solid waste compost quality produced in different cities of India in the perspective of developing quality control indices. *Waste Management, 30,* 192–201.

Sundberg, C., & Navia, T. (2014). Is there still a role for composting? *Waste Management & Research, 32,* 459–460.

Superintendencia de Servicios Públicos Domiciliarios. (2008). *Diagnóstico sectorial. Plantas de aprovechamiento de residuos sólidos.*

Thi, N., Kumar, G., & Lin, C. (2015). An overview of food waste management in developing countries: Current status and future perspective. *Journal of Environmental Management, 157,* 220–229.

Vélez, V. (2016). *Metodología para la caracterización de residuos sólidos en la cabecera de Versalles, Valle del Cauca,* Trabajo de grado del Programa de Ingeniería Sanitaria y Ambiental, Facultad de Ingeniería, Universidad del Valle, Cali, Colombia.

Yusti, E. (2017). *Evaluación económica de la gestión de residuos sólidos municipales en la zona urbana del municipio de Versalles -Valle del Cauca,* Trabajo de grado del Programa de Ingeniería Industrial, Facultad de Ingeniería, Universidad del Valle, Cali, Colombia.

Zurbrügg, C., Drescher, S., Rytz, I., Maqsood, S., & Enayetullah, I. (2005). Decentralised composting in Bangladesh a win-win situation for all stakeholders. *Resources, Conservation and Recycling, 43,* 281–292.

Organic Waste Composting through Nexus Thinking: Linking Soil and Waste as a Substantial Contribution to Sustainable Development

Hiroshan Hettiarachchi, Johan Bouma, Serena Caucci, and Lulu Zhang

Abstract This chapter explains why organic waste composting is con-sidered as one of the best examples to demonstrate the benefits of nexus thinking. Current literature is rich with information covering various aspects of composting process. However, it mainly represents two distinct fields: waste from the manage-ment point of view and soil/agriculture from the nutrient recycling point of view. It is hard to find information on how these two fields can benefit from each other, except for a few examples found within large agricultural fields/businesses. A pol-icy/institutional framework that supports a broader integration of management of such resources is lacking: a structure that goes beyond the typical municipal or ministerial boundaries. There is a clear need to address this gap, and nexus thinking can help immensely close the gap by facilitating the mindset needed for policy inte-gration. Good intention of being sustainable is not enough if there is no comprehen-sive plan to find a stable market for the compost as a product. Therefore, the chapter also discusses the strong need to have a good business case for composting projects. Composting can also support achieving the Sustainable Development Goals (SDGs) proposed by the United Nations. While directly supporting SDG 2 (*Zero hunger*), SDG 12 (*Responsible consumption and production*), and SDG 13 (*Climate action*), enhanced composting practices may also assist us reach several other targets speci-fied in other SDGs. While encouraging waste composting as a sustainable method of waste and soil management, we should also be cautious about the possible adverse effects compost can have on the environment and public health, especially due to some non-traditional raw materials that we use nowadays such as wastewater sludge and farm manure. Towards the end, we urge for the improvement of the entire chain ranging from waste generation to waste collection/separation to com-

H. Hettiarachchi (✉) · S. Caucci · L. Zhang
United Nations University (UNU-FLORES), Dresden, Sachsen, Germany
e-mail: hiroshanh@gmail.com; caucci@unu.edu; lzhang@unu.edu

J. Bouma
(Formerly) Wageningen University, Wageningen, The Netherlands
e-mail: johan.bouma@planet.nl

post formation and, finally, application to soil to ensure society receives the maximum benefit from composting.

Keywords Compost · Municipal solid waste (MSW) · Nexus thinking · Nutrients · Organic waste · Soil organic matter (SOM) · Sustainable Development Goals (SDGs) · Waste management

1 Background

Composting is a natural process of biological decomposition and stabilisation of organic waste (Oppliger and Duquenne 2016; Dollhofer and Zettl 2017). The nutrient-rich final product, that can be applied to land as soil fertiliser or stabiliser, offers significant benefits to agroecological systems as it combines environmental protection with sustainable agricultural production (Thanh et al. 2015; Román et al. 2015; Mbuligwe et al. 2002). The improvement of soil properties is a major benefit of compost application (Brändli et al. 2007).

Composting has also been gaining increasing attention as an alternative way of waste processing. In addition to the organic fraction of municipal solid waste (MSW), it is now also being adapted for treatment of various other types of organic waste such as farm manures, sewage sludge, and industrial sludge (Otoo and Drechsel 2018; Azim et al. 2018; Barker 1997). Indeed, the composting process has the ability to reduce pathogenic bacteria, viruses, and parasites in such waste material, which could otherwise pose a health risk. Although pathogens cannot be eliminated completely via composting, the presence of pathogens in compost is lower than in livestock manures (Wéry 2014).

In the above context, organic waste composting helps us to be more sustainable with how we manage our environmental resources. Nutrient recycling embedded in the concept of composting supports the idea of transitioning to a circular economy, which is currently being discussed in many international circles. There is also another international dialogue currently occurring on the developmental agenda put forward by the United Nations in 2015, which comprises of 17 goals to be achieved by 2030, i.e. Sustainable Development Goals (SDGs) (UN 2015a). As we discuss later in this chapter, organic waste composting directly addresses a few SDGs in addition to partially supporting several others.

Despite how interesting it sounds as an idea and the tremendous potential it offers from the circular economic point of view to achieve the SDGs, organic waste composting is not popular enough yet. Among many reasons, two stand out more prominently. One reason is the disconnect between the agricultural sector where compost is applied and the waste management sector where the bulk of raw material originates from. The policies and institutional structures we have today do not necessarily provide any space for the integrated management of resources (Hettiarachchi and Ardakanian 2016a). The other prominent reason is the lack of a strong business case. Many composting projects from around the world have failed due to the lack of the same (Hettiarachchi et al. 2018; Otoo and Drechsel 2018). Good intention of being sustainable itself is not enough if there is no comprehensive plan to find a stable market for the compost as a product. It is unfortunate that compost projects fail, while there is still a clear need and a reasonable demand for them in the agricultural sector. A policy framework that allows for broader integration of management of such resources, that goes beyond typical municipal or ministerial boundaries, is lacking (Hettiarachchi and Ardakanian 2016a; Subramanian et al. 2019).

The primary objective of this book is to shed light on the reasons behind the above gaps and introduce the potential of nexus thinking as a way forward. Nexus thinking facilitates the mindset needed for policy integration to promote the integrated management of resources (Hettiarachchi and Ardakanian 2016a; Subramanian et al. 2019). How we may bridge the gaps existing between soil and waste management through nexus thinking is elaborated in a later part of this chapter. Some emerging trends observed in the field of composting, both positive and negative, also demand attention. Among others, there is currently a growing concern about the spread of antibiotic resistance and its impact on humans and animals. The role played by the composting industry, in this regard, warrants a discussion. Compost made from municipal solid waste and poultry manure/litter may also contribute to the spread of pollutants of emerging concerns, including antimicrobial resistance determinants, micro- and nanoplastics, and organic pollutants.

In this context, the objective of this chapter is to introduce the entire book to set the scene for a deeper discussion. In the next sections, we present the above-mentioned information in greater detail.

2 Compost: A Sustainable Product for Soil Enrichment

There is an escalating concern about soil and land degradation, as global demands for food, water, energy, and raw materials continue to grow at unprecedented rates. The lack of success in meeting these demands can easily cause adverse ripple effects in other issues important for humanity such as hunger, poverty, peace, migration, and well-being. The Global Land Outlook (UNCCD 2017) estimates that one third of the global land surface is severely degraded, which has negative impacts on the well-being of at least 3.2 billion people across all continents. This results in losses of biodiversity and ecosystem services amounting to an economic loss of about 10% of the world's annual gross product in 2010. In the European Union, soil degradation causes an annual loss of 2.46 megagrams of soil per hectare each year (Mg/ha.year) from erosion-prone lands, equivalent to an area the size of Berlin to a depth of 1 m (Panagos and Borrelli 2017). Beyond Europe, conditions are worse: erosion rates in South America (3.53 Mg/ha.year), Africa (3.51 Mg/ha.year), and Asia

(3.47 Mg/ha.year) are even higher (Borrelli et al. 2017). In addition, low soil organic matter (SOM) is also becoming a concern. For example, 45% of European soils now have low SOM, mostly in Southern Europe, but also in areas of France, the UK, and Germany (Jones et al. 2012). This negatively affects soil quality as the water holding capacity is reduced, as are soil structural stability and the nutrient adsorption potential. To feed an ever-increasing population while adapting to climate change, protecting and restoring soils must be mandatory to maintain ecosystem services (Bastida et al. 2015).

There are several ways to overcome the long list of challenges from land degradation to soil losses and loss of biodiversity. One of the restoring approaches to recover and mitigate soil degradation is by preserving and enhancing SOM content in soils. The increase of SOM enhances the aggregation of soil particles, thus improving aggregate stability and soil structure (Apostolakis et al. 2017). Stable soil aggregates reduce the risks of soil surface sealing and associated surface run-off formation, minimise soil erosion, promote water infiltration, and enhance the soil water retention capacity (Martínez-Blanco et al. 2013; Lado et al. 2004). Additionally, increasing SOM mitigates global warming through sequestering carbon in soils and reducing greenhouse gas emission. It is important to state that the French proposal, suggesting an average increase of SOM by only 0.04% is now part of the Paris climate agreement of 2015 (Initiative 4 pour 1000 2018). Increased SOM also benefits the soil microbiome and biodiversity, which, in turn, facilitates better nutrient supply for plants and suppression of soilborne diseases (Cesarano et al. 2017; Martínez-Blanco et al. 2013).

The key question now is how we can increase SOM. Organic manure is widely used, but its quantity and quality are inadequate in many parts of the world. Green manuring – growing alfalfa and other crops that bind nitrogen through microbiological processes – is an option, but as a solution not in the least practical, because it does not commercially allow more attractive crops to be grown at the same time. Leaving crop residues on the land is another measure to increase SOM, but – as in the case of organic manure – its quantity is too limited. Even though there might not be one perfect solution, century-long experiences humanity had with composting suggest that compost might be the sustainable solution to increase SOM.

Compost is considered invaluable for soil scientists, land managers, and farmers, not necessarily in terms of monetary value, but due to the richness in composition with organic materials, multiple nutrients (e.g. N, P, K), microbiomes, and water that are vital for land-based production. This makes compost an even more attractive option for the purposes of land restoration. Some of these soil-focused benefits are highlighted in the case studies from Tanzania presented in Chaps. 5 and 8 and the one from Sri Lanka in Chap. 4. Yields of fields fertilised with compost were measured, but the associated soil processes, as a function of natural soil moisture regimes, were not identified which makes extrapolation of results difficult. From this perspective, soil scientists and land management specialists have a task to fulfil, i.e. to obtain optimal compost products that can restore soil and make use of the nutrients and water pool properties to generate specific environmental and socioeconomic benefits (Martínez-Blanco et al. 2013; Onwosi et al. 2017).

3   Composting: A Sustainable Method of Managing Organic Waste

Organic wastes commonly comprise food wastes, garden wastes, agricultural wastes, and some process residues. As the World Bank estimates, the per capita MSW generation has now reached 0.74 kg/day and is expected to increase by another 70% by 2050 (Kaza et al. 2018). It is important to understand that the largest fraction is almost always the organics, which can vary from 27% in OECD countries to 62% in East Asia and the Pacific (World Bank 2012). These numbers provide some perspective to the volume of organic waste that we produce and the difficulties we face in managing it. Undoubtedly, the effective management of organic solid waste is essential to mitigate environmental and socioeconomic risks of solid wastes.

From the management point of view, organic waste is an "issue" that we need to address timely, and there are standard engineering solutions such as landfilling or incineration that we can use. Organic waste is potentially an even bigger "issue" than the other types of waste, due to the damage it can cause to the environment and public health, if not managed properly (Hettiarachchi et al. 2018). However, from the sustainability point of view, organic waste is not an issue, rather it is a resource. Several alternative methods have been proposed to capitalise on the resource point of view of organic waste (Otoo and Drechsel 2018). Among these options, composting has gained wide attention because of low operating costs and high environmental compatibility (Onwosi et al. 2017). Some of these key aspects of composting as a waste management option are particularly highlighted in the two case studies from Colombia and India (Chaps. 2, 3, and 7). Moreover, in comparison to other approaches, composting is also attractive as it is one of the oldest, best-known, and well-established processes (Martínez-Blanco et al. 2013).

The impact compost makes on society goes far beyond its simplicity and meets many distinct objectives across sectors. For example, in MSW management, composting also helps to shrink the waste volume, which would otherwise occupy land fill space. From the viewpoint of resource recovery, which is an important component in the waste management pyramid (in importance order: waste prevention and minimisation, reusing, recycling, valorisation, and final disposal), the nutrient recycling aspect of composting makes it a very sustainable method of managing organic waste. From the environmental, economic, and social perspectives, composting turns organic wastes into a product (compost) useful for agricultural activities and soil restoration which is carbon-, water-, and nutrient-rich and free of most pathogens. Thus, nutrient recycling through composting addresses the well-known three pillars of sustainability: the society, the economy, and the environment.

4 Bridging the Gaps Through Nexus Thinking

The discussion above leads to one solid conclusion: composting is a sustainable method of managing organic waste and the final product – compost – is a sustainable way to enrich soils. Considering the large size of the global agricultural industry and the large volume of organic waste we produce, the two concepts should technically complement each other to create a very successful and large composting industry. However, we know that this is currently not the case. To address this, we need to look at the underlying reasons that have been briefly mentioned earlier: the disconnect between the soil and waste fields and the lack of a better business case.

The disconnect between the soil/agricultural sector and the waste sector is due to how the two sectors traditionally evolved independently. In general, MSW is handled by municipalities, while agricultural businesses are mainly in the private sector. Those who are involved in the two sectors are also trained professionally in two different settings. The same trend can be seen in the research and academia as well. Current literature on composting is rich with many journal articles, conference papers, and book chapters and also in the popular press, covering various aspects. However, these articles (and the authors) almost always represent only one side of the story: it is either about waste from the management point of view or about soil/agriculture from the nutrient recycling perspective. It is rare to find information on how these two sectors can benefit from each other.

Many composting businesses have suffered from lack of a better business case. If we ask a compost manufacturer, the answer would more likely be about the absence of a steady/profitable market for the product. It is true that compost making cannot survive as a business if the products cannot find a steady market. This is where most compost projects have failed in the past. If we ask farmers (in developing countries) why they are not interested in compost, the most probable answer would be that the mineral/chemical fertiliser is cheaper than compost due to government subsidies (ADB 2011). No one can expect farmers to give up on mineral/chemical fertilisers to buy more expensive compost, just because it is a more sustainable option. This implies the clash of different policies: while there are policies encouraging municipalities to be more sustainable in the ways they handle waste, other government policies have created fertiliser subsidies with the intention of helping farmers.

The quality of compost made from organic waste has also been raised as an issue. Implementing source separation is the simplest answer (Hettiarachchi et al. 2018). However, composting businesses are not able to implement any collection policies as it is the responsibility of the municipality. This tells us how important it is to have policies and collaborations that go beyond the traditional boundaries. Sectorial thinking does not help much as exhibited by the low volume of compost production compared to the sheer volume of organic waste we produce. Now the questions that remain are: who should take the lead in making compost out of organic waste? Is this an agricultural issue, or should it be up to the local municipality? Or else, should this be a topic for a wider discussion on environmental resources management? Any plan that involves both the agricultural and business communities in close interaction with the local and wider policy arena would certainly have the greatest potential to succeed.

Here, we can benefit from the emerging concept of nexus thinking in managing environmental resources. What nexus thinking promotes is a higher-level integration that goes beyond the disciplinary boundaries (Hettiarachchi and Ardakanian 2016a). One excellent example is wastewater recycling for agricultural irrigation. From the resource point of view, wastewater helps alleviate the supply issues faced by the water sector and water demand issues in the agricultural sector. This importance of wastewater recycling as a nexus example has been widely discussed before by UNU-FLORES in other publications (Hettiarachchi and Ardakanian 2016b, 2018). The topic of our current discussion – compost – is another very fitting example that can benefit from nexus thinking. Like the wastewater example, the compost chain too starts in the waste sector. But the final product feeds into the soil/agriculture and food sectors. Through its water retention properties, compost also contributes to the water sector by offering water savings in agricultural irrigation as well as higher purification of percolating water due to a higher soil adsorption capacity.

The above examples showcase how nexus thinking leads to a higher-level integration of environmental resources management. However, the beauty of the concept lies in its ability to force us to think through the policy infrastructure. Policy integration is a must for nexus thinking to be beneficial. It is not easy, but not impossible either, as shown in Chap. 2 with the Cajicá example. Through a participatory approach, the municipality of this small city in Colombia was able to involve all stakeholders to reinvent its policy structure to establish a successful composting programme in a few years. A similar approach is also showcased in Chap. 4 and in the Utilization of Organic Waste for Improvement of Agricultural Productivity (UOWIAP) Project from Accra, Ghana, presented in Chap. 6.

This illustrates an important emerging aspect of nexus thinking. Close cooperation and interaction among stakeholders involved in waste collection, compost formation, and its application in soils is essential. Interdisciplinarity brought by the stakeholders shows us how important it is to take a similar path in research, too, and conduct interdisciplinary research in close interaction with all stakeholders. This becomes increasingly important in the twenty-first century where information should be widely available and accessible and affects people's reaction to results and conclusions presented by scientists and policymakers alike (Bouma 2019a). Rather than studying "their" problems in a detached manner, it is more beneficial to tackle the problems thinking that they are "our" problems (Bouma 2018). Creating a feeling of ownership among stakeholders is crucial to achieve effects of practical significance.

5   Composting in the Sustainable Development Agenda

We are now being challenged by the implications of global changes. The population explosion that will result in an estimated ten billion people by the middle of the century (UN 2017), climate change, water scarcity, land degradation, and overall decline of environmental quality are some of the notable ones. Although we feel it now, the need for remedial action for the same challenges has long been recognised. In 1987 the iconic Brundtland Report on *Our Common Future* introduced the concept of sustainable development, emphasising the need to not only consider economic aspects but also social and environmental aspects when defining future actions aimed at maintaining a vital global environment (WCED 1987). The concept of sustainable development has, however, remained rather abstract in the international arena for a decade or so, until the Millennium Development Goals were unveiled in the year 2000 (UN 2015b). The concept has been further improved when the 17 Sustainable Development Goals (SDGs), that apply to the entire world, were defined in 2015. The SDGs do not present lofty, abstract goals as they have been formally adopted by 135 countries accepting the obligation to satisfy specific targets and indicators by 2030. Although SDGs are brought into discussions in Chaps. 5 and 6 of this book, it is apparent that the concept has not yet been completely internalised by the stakeholders. This is also evident from the fact that soil scientists have not been engaged with defining targets and indicators for the various SDGs (Bouma 2019a). Active engagement by all parties is important as we approach the year 2030 in only a decade from now.

Achieving the SDGs by 2030 is undoubtedly a very ambitious goal. The success of achieving them will surely depend not only on innovative solutions but also new thinking. It is also important to note the limitations we have in the traditional, individual scientific disciplines, in offering any comprehensive practical approaches that can result in achieving any of the SDGs (Bouma 2014; Keesstra et al. 2016). As previously discussed, nexus thinking, however, provides a platform for such practical solutions to prosper. Taking the resource perspective into account, the concept of integrated management of Water-Soil-Waste, put forward by UNU-FLORES, offers us a route to overcome some of these SDG challenges. Organic waste composting is an example highlighting the utility of nexus thinking to help us achieve the SDGs.

Enhanced composting efforts directly support the task of achieving SDG 2 (*Zero hunger*), SDG 12 (*Responsible consumption and production*), and SDG 13 (*Climate action*). Through its relevance to agriculture, nutrient recycling, and waste (converting a greenhouse gas emitter into a solution), the direct relevance of composting to above SDGs is readily understood. In addition, it will also help us partially address numerous other goals, which we might not instantly recognise the importance of until we look at the list of targets each goal carries. For example, SDG 6 (*Water and sanitation*) has a target (6.4) to increase water use efficiency. Compost, which is known for its ability to improve the water retention capacity of soils, can help us address this target from the agricultural water use efficiency point of view. SDG 11 (*Sustainable cities and communities*) has a target (11.6) about municipal and other waste management to minimise the adverse impact on the environment. SDG 15 (*Life on land*) is another example that has a target (15.3) to restore degraded land and soil. SDG 17 (*Partnerships for the goals*) encourages promoting effective public, public-private, and civil society partnerships in one target (17.17). Composting is one good example where such partnerships already function very well, especially when the MSW management lies in the public sector and the usage of compost mainly happens in the private sector, while compost is currently made by organisations representing all of the above sectors.

As mentioned above, compost offers a unique and adequate source of carbon to increase SOM. However, the chain from waste generation, compost preparation, and application to soils is long and complex and involves many actors, each of which often has contradicting interests and demands. New approaches always offer resistance, in this case, traditional land users who find it hard to change existing practices. Composting has a problem in that it has a negative connotation to many people, which is strengthened when animal manure (or even human excreta in some cases) with pathogens and industrial waste with heavy metals are part of the process. How do we overcome these often psychological barriers in integrating compost into agricultural production systems? So far, we have discussed how composting can help us achieve the SDGs. In fact, the SDGs themselves can be used in return to overcome psychological barriers discussed above and make compost popular as a solution. There are many awareness-raising campaigns about the SDGs, their impact, and how to achieve them. Such campaigns could, when well-designed, become effective information channels serving the composting industry.

Recently, the concept of soil security has also been advanced in soil science research (Field et al. 2017) based on the 5Cs: Condition (actual condition of the soil), Capability (what can be achieved with improved management), Capital (how the soil compares with others), Connectivity (connection with stakeholders and the policy arena), and Codification (role of soils in laws and regulations). There is a clear link between soil security and not only the SDGs (Bouma 2019b) but also with the Water-Soil-Waste Nexus as discussed in this book. The soil security concept puts the soil in a wider societal context, as also expected from the integrated management of Water-Soil-Waste.

6   Emerging Trends: New Opportunities Versus New Challenges

New trends related to composting have been recognised lately: some are positive, and others might be alarming. Positive trends include the small but relatively steady demand for compost caused by the popularity of organic farming (USADA 2011), increased awareness of the role of waste in a circular economy (Hettiarachchi 2018), push for use of crop residue for composting rather than burning (Subramanian et al. 2019), and increased use of non-traditional raw material to make new compost products through various co-composting technologies (as described in Chaps. 9 and 10 in the book). The use of various non-traditional raw materials, seemingly a positive trend, might have inadvertently opened a Pandora's box for the composting industry from the public health perspective as further described in the case study from Sri Lanka in Chap. 4. While encouraging introducing organic waste composting as a sustainable waste and soil management practice, we should also be alert and cautious about the possible adverse impacts it can cause. These limitations should be understood and respected, and care should be taken to identify the threats scientifically to avoid them at all costs.

While being a source of valuable plant nutrients, compost could also be a source of a broad array of contaminants of emerging concern (CECs) which are not completely understood or removed during the composting process (Watteau et al. 2018; European Commission 2003). This group of contaminants includes a variety of substances that are commonly used in daily life such as pharmaceuticals and their breakdown metabolites. Antibiotic-resistant bacteria/genes have also been added to this list recently. CECs are introduced to the environment mainly through wastewater treatment plants, activated sludge, and manure and cause known or suspected adverse ecological and human health effects (Snow et al. 2017; Zuloaga et al. 2012). The agricultural use of sewage sludge/compost and manure as a fertiliser has become one of the most widespread routes for the release of these substances in the environment since it provides an opportunity to recycle plant nutrients and organic matter to the soil for crop production (Zuloaga et al. 2012). In this way, these contaminants can enter the food chain via crop plants and compromise human, animal, and environmental health (Wiechmann et al. 2015).

Plastics and microplastics are also now considered within the category of CECs due to their widespread presence, persistence, and multiple ecotoxicological and ecological hazards that these particles pose to the terrestrial ecosystem (Souza Machado et al. 2018; Avio et al. 2017). Agroecosystems are particularly emphasised as the major entry point for microplastics in the environment via land application of sludge and compost (Rillig 2012). Pharmaceuticals and especially antibiotics are the other groups in CECs which have the potential to enter the environment and cause known or suspected adverse ecological or human health effects (Snow et al. 2017). Globally, antibiotics have drawn considerable attention due to their increasing use and capacity for being a selective pressure for multi-resistant bacteria (Grenni et al. 2018; Caucci et al. 2016). Antibiotics have the potential to affect natural microbial communities after their entry into the environment (Grenni et al. 2018). Like microplastics, antibiotic residues are present in the environment for a while, but the impacts of antibiotics and antibiotic-resistant genes of bacteria are not well-known yet (Bengtsson-Palme et al. 2018; Snow et al. 2017).

The severity and extent of adverse environmental and health impacts posed by these contaminants are generally unknown due to their complex interactions and transformation processes. Thus, the quality of compost has become a subjective term, while the legal framework and standards differ vastly between countries (Hogg et al. 2002). While there are ongoing efforts on improving quality standards for composting, as it can be seen in the case of European Union which now has a comprehensive regulatory framework for different types of fertilisers (discussed in case study from Italy in Chap. 9 and from Finland in Chap. 10), it is not the case in many other countries around the world (Pollak and Favoino 2004; European Compost Network 2014). These emerging issues need to be further discussed among policymakers and scientists in the scope of sustainability of organic waste recycling and its application to reflect the current level of knowledge on CECs in regulatory frameworks and quality standards to minimise any potential risk.

7   The Way Forward

Compost presents a good example of considering organic waste as a resource rather than a nuisance to be discarded. It is a resource currently "out of place" that needs to be "put in place" for the benefit of the society, the economy, and the environment – the three pillars of sustainability. It provides an opportunity to realise an attractive circular approach which is now widely emphasised for a more efficient use of our limited natural resources. Rather than throwing the organic waste away, we can close the cycle by returning part of it to the land, where it came from.

Increased soil quality, evidenced by higher productivity and soil resilience, presents the most effective testimony to the societal significance of compost. The chapters of this book show that many operational techniques are available now to produce a range of composts of different composition. Citizen participation in collecting and separating waste has been proven effective in several countries. One such example is the Cajicá case in Colombia as presented in Chap. 2. However, the continuation of such systems is all too often subject to local politics and unreliable funding, and this may result in unsustainable systems.

Therefore, the question should be raised as to how this vulnerability of the system can be reduced by establishing a healthy economic basis for the complete compost chain. This can be achieved by demonstrating that the nutrients in compost and increased SOM are economically and environmentally attractive as they produce not only higher but also more reliable crop yields while improving soil biodiversity. This can best be achieved by well-documented case studies, several of which are presented in the chapters of this book. In-depth soil studies of interrelated physical, chemical, and biological processes are quite limited so far, and this needs more attention in the future. It is important to consider the entire chain ranging from waste generation to waste collection/separation to compost formation and finally application to soil. Unfortunately, the emphasis is so far on the first part of the chain: waste collection and compost formation.

We should also recognise some of the operational problems inherent in compost applications such as time lag: additions of compost to soils do deliver nutrients but do not instantly increase the SOM. Continued soil biological activity is needed to incorporate the new organic matter into the soil fabric. This process can take at least several years. Classic research, applying compost in different quantities to different fields and measuring yields, as elaborated in Chap. 5, does, therefore, not provide an answer to the attractive long-term perspective of adding compost to soils. This tells us again that we should also invest in field studies, finding examples of land where compost has already been added for long periods, comparing results with similar data for soils where compost has not been applied. Only through this can we show relevant differences. While doing so, we should realise that different soils have different chemical properties and soil moisture regimes, leading to results that will always be site-specific.

The successful application of compost, leading to higher agricultural production and increased soil quality, is a very positive result that farmers can appreciate. Meeting the needs of farmers supports the societal aspect of sustainable development. Applying compost with nutrients means less usage of chemical fertilisers, and this implies possible savings. Such results should convince even the most critical compost users, because usually "money talks" most convincingly.

Once compost is seen by farmers as a crucial element of future soil management, the demand for compost and the demand for organic waste will also increase. When the end users in the agricultural sector, i.e. farmers, become prosperous, the social benefits in the compost chain would eventually take the reverse path, due to an increased demand ensuring that those who work in the waste sector are also economically and socially benefiting from the process. The reverse path can also be beneficial from the capacity development point of view: scientific evidence not only shows that using compost is economically attractive for farmers but also encourages farmers who are already convinced to convince others. Finally, the policy arena should also be addressed making sure that regulatory measures support rather than suppress the proper application of compost, as discussed in Chap. 5 and several others.

References

ADB. (2011). *Toward sustainable municipal organic waste management in South Asia: A guidebook for policy makers and practitioners*. Mandaluyong City: Asian Development Bank.

Apostolakis, A., Panakoulia, S., Nikolaidis, N. P., & Paranychianakis, N. V. (2017). Shifts in soil structure and soil organic matter in a chronosequence of set-aside fields. *Soil and Tillage Research, 174*, 113–119.

Avio, C. G., Gorbi, S., & Regoli, F. (2017). Plastics and microplastics in the oceans: From emerging pollutants to emerged threat. *Marine Environmental Research, 128*, 2–11.

Azim, K., Soudi, B., Boukhari, S., Perissol, C., Roussos, S., & Thami Alami, I. (2018). Composting parameters and compost quality: A literature review. *Organic Agriculture, 8*(2), 141–158.

Barker, A. V. (1997). Composition and uses of compost. In *Agricultural uses of by-products and wastes, American chemical society 668* (pp. 140–162). Washington, DC: American Chemical Society.

Bastida, F., Selevsek, N., Torres, I. F., Hernández, T., & García, C. (2015). Soil restoration with organic amendments: Linking cellular functionality and ecosystem processes. *Scientific Reports, 5*, 15550.

Bengtsson-Palme, J., Kristiansson, E., & Larsson, J. D. G. (2018). Environmental factors influencing the development and spread of antibiotic resistance. *FEMS Microbiology Reviews, 42*(1), 68–80.

Borrelli, P., Robinson, D. A., Fleischer, L. R., Lugato, E., Ballabio, C., Alewell, C., Meusburger, K., Modugno, S., Schütt, B., Ferro, V., & Bagarell, V. (2017). An assessment of the global impact of 21st century land use change on soil erosion. *Nature Communications, 8*(1), 2013.

Bouma, J. (2014). Soil science contributions towards sustainable development goals and their implementation: Linking soil functions with ecosystem services. *Journal of Plant Nutrition and Soil Science, 177*(2), 111–120.

Bouma, J. (2018). The challenge of soil science meeting society's demands in a "post-truth", "fact-free" world. *Geoderma, 310,* 22–28. https://doi.org/10.1016/j.geoderma.2017.09.017. Viewed in May 2019.

Bouma, J. (2019a). How to communicate soil expertise more effectively in the information age when aiming at the UN sustainable development goals. *Sand Use and Management, 35*(1), 32–38.

Bouma, J. (2019b). Soil security in sustainable development. *Soil Systems, 3,* 5.

Brändli, R. C., Kupper, T., Bucheli, T. D., Zennegg, M., Huber, S., Ortelli, D., Müller, J., et al. (2007). Organic pollutants in compost and Digestate: Part 2. Polychlorinated dibenzo-p-dioxins, and -furans, dioxin-like polychlorinated biphenyls, brominated flame retardants, perfluorinated alkyl substances, pesticides, and other compounds. *Journal of Environmental Monitoring, 9*(5), 465–472.

Caucci, S., Karkman, A., Cacace, D., Rybicki, M., Timpel, P., Voolaid, V., Gurke, R., Virta, M., & Berendonk, T. U. (2016). Seasonality of antibiotic prescriptions for outpatients and resistance genes in sewers and wastewater treatment plant outflow. *FEMS Microbiology Ecology, 92,* fiw060.

Cesarano, G., De Filippis, F., La Storia, A., Scala, F., & Bonanomi, G. (2017). Organic amendment type and application frequency affect crop yields, soil fertility and microbiome composition. *Applied Soil Ecology., 120,* 254–264.

Dollhofer, M., & Zettl, E. (2017). *Quality assurance of compost and digestate experiences from Germany.* German Environment Agency.

European Commission. (2003). Directorate-general for the environment, and applying compost: Benefits and needs. In *Seminar proceedings, Brussels, 22–23 November 2001.* Vienna: Federal Ministry of Agriculture, Forestry, Environment and Water Management.

European Compost Network. (2014). *European quality assurance scheme for compost and digestate ECN-QAS.*

Field, D. J., Morgan, C. I. S., & Mc Bratney, A. D. (2017). *Global soil security.* Cham: Springer Publishing.

Grenni, P., Ancona, V., & Caracciolo, A. B. (2018). Ecological effects of antibiotics on natural ecosystems: A review. *Microchemical Journal, 136,* 25–39.

Hettiarachchi, H. (2018). Going full circle: Why recycling isn't enough. *The Japan Times.* https://www.japantimes.co.jp/opinion/2018/10/15/commentary/japan-commentary/going-full-circle-recycling-isnt-enough/#.XOaOZYgzY2w. Viewed in May 2019.

Hettiarachchi, H., & Ardakanian, R. (2016a). *Environmental resource management and Nexus approach: Managing water, soil, and waste in the context of global change.* Basel: Springer Nature.

Hettiarachchi, H., & Ardakanian, R. (2016b). *Good practice examples of wastewater reuse.* Dresden: UNU-FLORES.

Hettiarachchi, H., & Ardakanian, R. (2018). *Safe use of wastewater in agriculture: From concept to implementation.* Cham: Springer Nature.

Hettiarachchi, H., Meegoda, J. N., & Ryu, S. (2018). Organic waste buyback as a viable method to enhance sustainable municipal solid waste management in developing countries. *International Journal of Environmental Research and Public Health, 15*(11), 2483.

Hogg, D., Barth, J., & Favoino, E. (2002). *Comparison of compost standards within the EU, North America and Australasia.* The Waste and Resources Action Programme.

Initiative 4 pour 1000. (2018). What is the "4 per 1000" Initiative? Initiative « 4 pour 1000 » – ecrétariat Exécutif c/o CGIAR System Organization, 1000 av. Agropolis, 34394 Montpellier, France.

Jones, A., et al. (2012). The state of soil in Europe. In *A contribution of the JRC to the European Environment Agency's environment state and outlook report.* Luxembourg: Publications Office of the European Union.

Kaza, S., Yao, L., Bhada-Tata, P., & Van Woerden, F. (2018). *What a waste 2.0: A global snapshot of solid waste management to 2050.* Washington, DC: World Bank Publications.

Keesstra, S. D., Bouma, J., Wallinga, J., Tittonell, P., Smith, P., Cerda, A., Montanarella, L., Quinton, J., Pachepsky, Y., van der Putten, W. H., Bardgett, R. D., Moolenaar, S., Mol, G., & Fresco, L. O. (2016). The significance of soils and soil science towards realization of the United Nations sustainable development goals. *The Soil, 2,* 111–128.

Lado, M., Paz, A., & Ben-Hur, M. (2004). Organic matter and aggregate size interactions in infiltration, seal formation, and soil loss. *Soil Science Society of America Journal., 68*(3), 935–942.

Martínez-Blanco, J., Lazcano, C., Christensen, T. H., Muñoz, P., Rieradevall, J., Møller, J., Antón, A., & Boldrin, A. (2013). Compost benefits for agriculture evaluated by life cycle assessment: A review. *Agronomy for Sustainable Development, 33*(4), 721–732.

Mbuligwe, S. E., Kassenga, G. R., Kaseva, M. E., & Chaggu, E. J. (2002). Potential and constraints of composting domestic solid waste in developing countries: Findings from a pilot study in Dar Es Salaam, Tanzania. *Resources, Conservation and Recycling, 36*(1), 45–59.

Onwosi, C. O., Igbokwe, V. C., Odimba, J. N., Eke, I. E., Nwankwoala, M. O., Iroh, I. N., & Ezeogu, L. I. (2017). Composting technology in waste stabilization: On the methods, challenges and future prospects. *Journal of Environmental Management, 190,* 140–157.

Oppliger, A., & Duquenne, P. (2016). Highly contaminated workplaces. In *Environmental mycology in public health* (pp. 79–105).

Otoo, M., & Drechsel, P. (2018). *Resource recovery from waste: Business models for energy, nutrient and water reuse in low- and middle-income countries.* Oxon: Routledge – Earthscan.

Panagos, P., & Borrelli, P. (2017). Soil erosion in Europe: Current status, challenges and future developments. In *All that soil erosion: The global task to conserve our soil resources* (pp. 20–21). Soil Environment Center of the Korea.

Pollak, M., & Favoino, E. (2004). *Heavy metals and organic compounds from wastes used as organic fertilisers.* Directorate-General for the Environment of the European Commission.

Rillig, M. C. (2012). Microplastic in terrestrial ecosystems and the soil? *Environmental Science & Technology, 46*(12), 6453–6454.

Román, P., Martínez, M. M., & Pantoja, A. (2015). *Farmer's compost handbook experiences in Latin America.* Santiago: Food and Agriculture Organization of the United Nations (FAO).

Snow, D. D., Cassada, D. A., Larsen, M. L., Mware, N. A., Li, X., D'Alessio, M., Zhang, Y., & Sallach, J. B. (2017). Detection, occurrence and fate of emerging contaminants in agricultural environments. *Water Environment Research, 89*(10), 897–920.

Souza, M., de Anderson, A., Kloas, W., Zarfl, C., Hempel, S., & Rillig, M. C. (2018). Microplastics as an emerging threat to terrestrial ecosystems. *Global Change Biology, 24*(4), 1405–1416.

Subramanian, B., Hettiarachchi, H., & Meegoda, J. N. (2019). Crop residue burning in India: Policy challenges and potential solutions. *International Journal of Environmental Research and Public Health, 16,* 832.

Thanh, T. H., Yabar, H., & Higano, Y. (2015). Analysis of the environmental benefits of introducing municipal organic waste recovery in Hanoi City, Vietnam. *Procedia Environmental Sciences, 28,* 185–194.

UN. (2015a). *Transforming our world: the 2030 Agenda for Sustainable Development.* The resolution adopted by the General Assembly on 25 September 2015. United Nations.

UN. (2015b). *The Millennium Development Goals report 2015.* New York: United Nations.

UN. (2017). *World population prospects: The 2017 revision.* United Nations Department of Economic and Social Affairs – Population Division.

UNCCD. (2017). *The global land outlook* (1st ed.). Bonn: United Nations Convention to Combat Desertification.

USADA. (2011). Guidance: Compost and vermicompost in organic crop production. In *National organic program*. Washington, DC: United States Department of Agriculture.

Watteau, F., Dignac, M. F., Bouchard, A., Revallier, A., & Houot, S. (2018). Microplastic detection in soil amended with municipal solid waste composts as revealed by transmission electronic microscopy and pyrolysis/GC/MS. *Frontiers in Sustainable Food Systems, 2*.

WCED. (1987). Our common future. In *World commission on environment and development*. Oxford: Oxford University Press.

Wéry, N. (2014). Bioaerosols from composting facilities a review. *Frontiers in Cellular and Infection Microbiology, 4*.

Wiechmann, B., Dienemann, C., Kabbe, C., Brandt, S., Vogel, I., & Roskosch, A. (2015). *Sewage sludge management in Germany*. German Federal Environmental Agency.

World Bank. (2012). WHAT a WASTE: A global review of solid waste management. In *Urban development series, knowledge papers; urban development & local government unit*. Washington, DC: World Bank.

Zuloaga, O., Navarro, P., Bizkarguenaga, E., Iparraguirre, A., Vallejo, A., Olivares, M., & Prieto, A. (2012, July). Overview of extraction, clean-up and detection techniques for the determination of organic pollutants in sewage sludge: A review. *Analytica Chimica Acta, 736*, 7–29.

5

Composting as a Municipal Solid Waste Management Strategy: Lessons Learned from Cajicá, Colombia

Cristian Rivera Machado and Hiroshan Hettiarachchi

Abstract Municipal solid waste (MSW) generated in developing countries usually contains a high percentage of organic material. When not properly managed, organic waste is known for creating many environmental issues. Greenhouse gas (GHG) emissions, soil and water contamination, and air pollution are a few examples. On the other hand, proper and sustainable management of organic waste can not only bring economic gains but also reduce the waste volume that is sent for final disposal. Composting is one such recovery method, in which the end product – compost – eventually helps the agricultural industry, and other sectors, making the process an excellent example of nexus thinking in integrated management of environmental resources. The aim of this chapter is to discuss how Cajicá, a small city in Colombia, approached this issue in a methodical way to eventually became one of the leading organic waste composting examples in the whole world, as recognised by the United Nations Environment Programme in 2017. Cajicá launched a source separation and composting initiative called Green Containers Program (GCP) in 2008, based on a successful pilot project conducted in 2005. The organic waste separated at source collected from households, commercial entities, schools, and universities are brought to a privately operated composting plant chosen by the city to produce compost. The compost plant sells compost to the agricultural sector. The participants in the GCP could also receive a bag of compost every 2 months as a token of appreciation. The Cajicá case presents us with many lessons of good practice, not only in the sustainable management of waste but also in stakeholder engagement. It specifically shows how stakeholders should be brought together for long-lasting collaboration and the benefits to society. Finding the correct business model for the project, efforts made in educating the future generation, and technology adaptation to local conditions are also seen as positive experiences that others can learn from in the case of Cajicá's GCP. Some of the concerns and potential threats observed include the high dependency GCP has on two institutions: the programme financially depends completely on the municipality, and the composting operation

C. R. Machado · H. Hettiarachchi (✉)
United Nations University (UNU-FLORES), Dresden, Sachsen, Germany
e-mail: hiroshanh@gmail.com

depends completely on one private facility. GCP will benefit from having contingency plans to reduce the risk of having these high dependencies.

Keywords Cajicá Municipality · Colombia · Compost · Municipal solid waste (MSW) · Nexus thinking · Organic waste · Waste management

1 Introduction

Organic waste is known for posing a wide range of environmental challenges such as leachate production, greenhouse gas (GHG) emissions, offensive odours, and soil/water contamination, when they are disposed in landfills or especially uncontrolled dump sites (Hettiarachchi et al. 2018a). On the other hand, the same can be put to good use by society, if managed properly using sustainable recovery alternatives such as composting, anaerobic digestion, and thermal treatment. Such techniques can help us significantly reduce the organic waste that goes for final disposal in landfills and dump sites (Pace et al. 2018). These alternatives also offer other environmental benefits such as reduction of GHG emissions and leachate production in landfills (USEPA 2016; Adhikari et al. 2009). When used as soil amendments, the composting option also provides nutrient recycling, increased water retention capacity, and improved soil structure (Wei et al. 2017).

Municipal solid waste (MSW) produced in low- and middle-income countries usually has a much higher organic fraction compared to the same volume in high-income countries (Kaza et al. 2018). This is also true for the Latin American and Caribbean (LAC) region where 52% of the MSW produced is organic. High organic content means that there is room for recovery such as making compost or biogas production (Hettiarachchi et al. 2018a, b). Since waste separation is not a common practice in the region, they miss an opportunity for resource recovery as well as generation of some cash from such recovery activities. However, there is one community in LAC that went against all odds to make composting as a part of their municipality's MSW management strategy and finally became a world-class example in 10 years. This community is in Cajicá, Colombia.

Cajicá Municipality started organic waste composting as a pilot project in 2005. Since the success depends heavily on source separation, they put a lot of efforts in awareness-raising campaigns. With positive results from the pilot, finally they launched the full programme in 2008 which was eventually called the Green Containers Program (GCP). Now, after 10 years, they have accomplished much success. In 2017, UN Environment selected five cities from around the world to highlight new experiences with a strong approach to waste management (UN Environment 2017). Cajicá (Colombia) was among these five cities together with Osaka (Japan), Alappuzha (India), Ljubljana (Slovenia), and Penang (Malaysia). In the case of

Cajicá, it was selected mainly because of the Green Containers Program launched by the Cajicá Municipality about a decade ago.

This manuscript presents the case of Cajicá starting from 2005 to the present situation. In addition, a review of MSW and MSW management in Colombia is also presented prior to that from the material as well as policy perspectives. The Cajicá example is then presented considering the institutional, educational, social, technological, environmental, and financial aspects of the scheme. Such information may be particularly useful for other municipalities, especially in developing countries, to improve their performance in waste management.

2 Waste Management in Colombian Municipalities

Colombia generates about 32,000 tons of MSW daily, out of which 85% is disposed or treated, while the rest goes into recycling (DNP 2016). Landfilling is now the most widely used technique for final disposal throughout the country, and in 2016, almost 11 million tons of MSW was disposed in Colombian landfills. The MSW collected by Colombian municipalities usually presents a consistently high percentage of organic waste. For instance, Bogotá generated 57% organic waste, 13.25% plastic, 8.92% paper and paperboard, 1.91% textiles, 1.30% wood, 1.12% metals, 2.76% glass, and 1.99% other wastes (Alcaldía de Bogotá 2015). The MSW composition in Colombian municipalities is somewhat homogeneous as evident in Fig.. With this high organic content in MSW, much of it can be valorised through composting or biogas production.

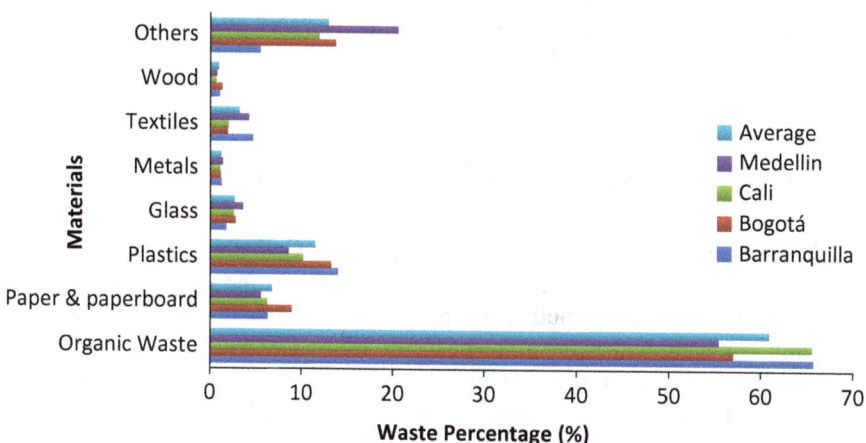

Fig. Waste characterisation and comparison in the principal cities of Colombia. (Source: Alcaldía de Barranquilla 2015; Alcaldía de Bogotá 2015; Alcaldía de Medellín 2015; Alcaldía de Santiago de Cali 2015)

Between 2005 and 2014, the national collection rate increased from 75.8% to 80%, and urban collection rate increased from 94.6% to 97.8%. At the national level, about 73% is served by private vendors that have financial capacity to provide a reasonable sanitary service (Banco Mundial 2015). However, there are still deficiencies in infrastructure and institutional capacities to serve rural areas. Waste collection coverage in rural Colombia is at about 24% (DANE 2013).

Until the 1980s and 1990s, open dumps were the predominant waste disposal facilities in Colombia. However, with a mixture of regulations, sanctions, and financial instruments introduced in the recent past, disposal in the open dumps is finally coming to an end, though gradually (DNP 2016). In addition, Colombia has also made progress with the application of better business models in the sanitary services sector, which has also resulted in increased waste collection rates in the urban areas and better quality in the final disposal methods and facilities (DNP 2016).

Landfilling is the final disposal method recommended in the national regulation, and currently 891 municipalities use landfills. Another 110 municipalities continue to practise open dumping, burying, burning, and disposing in waterbodies (Superintendence of Public Services 2017). The inadequate environmental performance of some of the landfills is also a major concern. In 2011, at least 30% of all the landfills in the country did not comply with environmental standards and regulation (OECD/UN ECLAC 2014).

Waste recovery in Colombia is mainly limited to the recycling of paper, cardboard, plastics, and metals. Other sustainable recovery methods pertaining to organic waste such as nutrient recovery through composting or energy recovery through biogas production have not become well recognised yet. Recycling is conducted by a workforce engaged in collection: 13,984 organised waste pickers and another 12,000 in the informal sector. In Bogotá alone, around 14,000 waste pickers depend on formal and informal recycling as their livelihood. Approximately 55% of the municipal-level recycling is the result of the informal sector (OECD/UN ECLAC 2014) although scavenging on landfills/dump sites was prohibited through the National Decree 1713 of 2002.

2.1 Policy Support Received by Municipalities

The institutional framework for MSW management in Colombia is characterised by multiple actors that can be clustered into three tiers. At the national level, three ministries and a national authority are involved – Health, Environment, and Housing – as well as the Superintendence of Household Public Services. At the regional level, the environmental authorities are in charge of regulation enforcement and monitoring. At the local level, the municipalities are in charge of the implementation of the MSW management policies.

The policy framework of MSW management in Colombia has evolved much during the past two decades. The basic structure was established in 1994 through Law 142 on Household Public Services. In 1998, the National Policy of Integrated

Waste Management emphasised on the objectives to minimise the generation of waste, increase recovery, and use landfills for final disposal. In 2002, Decree 1713 defined the concepts related to waste management and the requirements for the management of MSW, regarding the collection, recycling, and disposal (OECD/UN ECLAC 2014).

Based on the Regulation 1045 of 2003, every municipality is required to develop an Integrated Solid Waste Management Plan or more popularly known by its Spanish acronym PGIRS. This plan (PGIRS) defines the mechanisms, responsibilities, and methodologies for the municipalities to become guarantors of MSW management in their jurisdictions. PGIRS developed based on Regulation 1045 of 2003 was later found to be rather weak for practical formulation and implementation (Ministry of Housing, Cities and Territories 2014). As a result, the Ministry of Environment, together with the Ministry of Housing, issued a reformulated methodology – Regulation 754 of 2014. It consisted a new approach and contained features that could promote better ways to design and implement PGIRS in the entire country. PGIRS is perhaps the most important municipal waste management regulation introduced thus far in Colombia.

The National Decree 1077 of 2015 (before Decree 2981 of 2013) describes the municipal-level sanitary services for the entire country. These activities are (1) collection, (2) transport, (3) sweeping and cleaning of roads and public areas, (4) washing of public areas, (5) lawn mowing and tree pruning in public areas, (6) transfer, (7) treatment, (8) recovery and valorisation (without energy recovery), and (9) final disposal (Ministry of Housing, Cities and Territory 2015). The tariff framework included in Regulation 720 of 2015 (issued by the Regulatory Commission for Drinking Water and Basic Sanitation (CRA)) has fixed ceiling prices defined for waste management and sanitary services.

Waste recovery is yet to receive the attention of the policy sector. As mentioned, recycling has been carried out significantly by waste pickers. Thanks to some progressive steps made in the 1990s such as the creation of the Ministry of Environment, several regulations were introduced later recognising the role of waste pickers and identifying them as a vulnerable community (Red de ciudades cómo vamos 2014). However, only until the introduction of a new policy on waste management, CONPES 3874 in 2016, which enabled some nutrient recovery activities such as composting to foster.

2.2   Cajicá Municipality, Colombia

Cajicá is a municipality located in the Bogotá metropolitan area. It is positioned in the north of the Capital District, 39 km from Bogotá, at an altitude of 2598 m above sea level. It is surrounded by the municipalities of Zipaquirá to the north, Chia to the south, Tabio to the west, and Sopó to the east (Alcaldía Municipal de Cajicá 2018). As per the Colombian National Census conducted, Cajicá's population in year 2005 was 44,721. The projections made based on the last survey estimates the population for 2019 to be 61,549 with a distribution of 63% in urban settings and the rest in the rural areas (DANE 2005). The Sanitary Collection Enterprise (EPC) (Spanish acronym) is the entity in charge of Cajicá's sanitary services, under the authority of the municipality, and mainly owned by the municipality. Their service covers different categories of customers such as households, industries, and commercial entities. In the case of households, they are further classified into six categories called Socioeconomic Classes (SECs) based on income (the lowest-income households in SEC 1 and the highest in SEC 6). SEC statistics for the period 2010–2014 are presented in Fig. In 2014, the coverage of sanitary collection service for the Cajicá urban areas reached almost 100% (DANE 2014).

MSW Generation and Composition in Cajicá The waste generation rate in Cajicá has been continuously increasing during the past decade. Increase in population, growth of the middle class, changes in consumption habits, and urban expansion are among the main reasons. Table presents the daily average MSW generation data per capita for Cajicá from 2009 to 2014 (Alcaldía Municipal de Cajicá 2015).

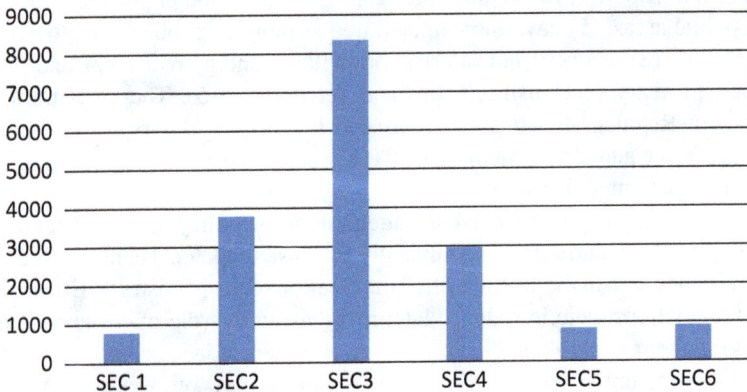

Fig. Customers of sanitary service in 2015 per SEC criteria. (Source: Alcaldía Municipal de Cajicá 2015)

Table Per capita MSW generation in Cajicá from 2009 to 2014	Year	MSW generation per capita (kg/capita/day)
	2009	0.58
	2010	0.62
	2011	0.66
	2012	0.72
	2013	0.78
	2014	0.84

Source: Alcaldía Municipal de Cajicá (2015)

Figure presents the composition of MSW that reached the final disposal from Cajicá. The low fraction of organic waste in it is clearly visible. This is due to the organic waste composting programme currently run by the municipality of Cajicá, which is the focus of this manuscript.

Final Disposal of MSW in Cajicá "Nuevo Mondoñedo" landfill is the facility used for final disposal of MSW collected in Cajicá. It is a landfill that serves 80 municipalities in the region, located in Bojacá which is about 50 km from Cajicá. The amount of MSW disposed has been increasing in the recent years according to data presented in Fig. The sudden drop in waste disposal between 2008 and 2009 is believed to be due to the diversion of organic fraction from landfill to sustain

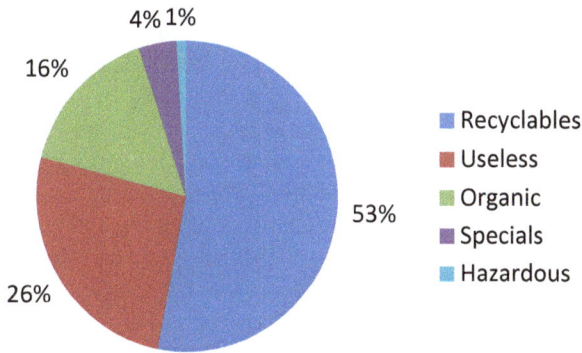

Fig. Cajicá's waste characterisation in the landfill in 2014. (Data: Alcaldía Municipal de Cajicá 2015)

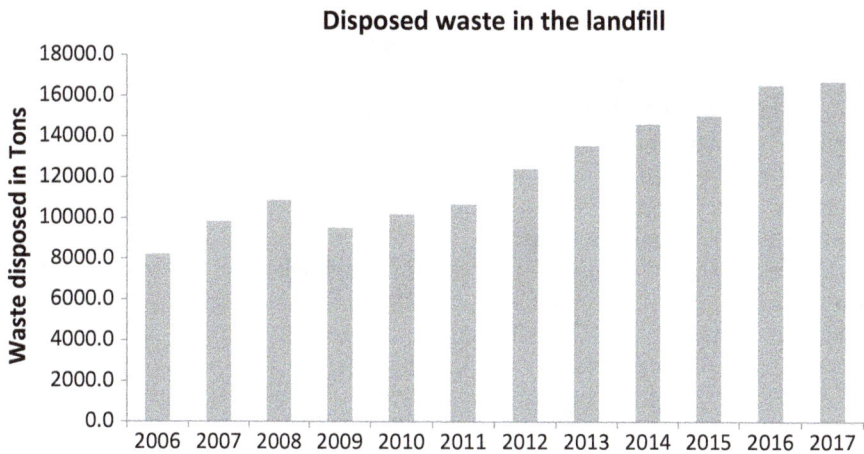

Fig. Cajicá's disposed waste between 2006 and 2014. (Data: Alcaldía Municipal de Cajicá 2015, and information given personally in Cajicá by EPC in November 2018)

the composting programme, as the timing coincides with the start of the Green Containers Program (Alcaldía Municipal de Cajicá 2015).

3 Green Containers Program: The Composting Initiative Launched by the Cajicá Municipality

The organic waste recovery efforts in Cajicá began in 2005 as a small composting pilot project conducted by the municipality. Cajicá Municipality, EPC (which was introduced earlier), and some motivated residents initiated a small organic waste source separation campaign to feed the composting pilot project. They collected 2 tons of organic waste through composting and vermiculture and converted compost during the pilot, which was distributed among the participating families (Ospina 2009). The pilot was a significant experience that allowed the Cajicá Municipality to gather convincing data, experience, and arguments to scale up the project to reach its status today.

Based on the above experience, Cajicá Municipality officially launched a source separation and citizen awareness programme mainly focusing on organic waste management through the Decree 061 of 2005. The programme aimed to increase the awareness of citizens in source separation for composting and recycling and instil good practices of sustainable waste management (Alcaldía Municipal de Cajicá 2015). The programme was planned and developed between 2006 and 2008 with the involvement of the following stakeholders from the region: Cajicá Municipality, the public, EPC, waste facility managers, government environmental authority, and industries. Then, at the end of 2008, the source separation programme called Green Containers Program (GCP, hereafter) was born, initially focusing only on some selected neighbourhoods.

Cajicá's Municipality began to distribute green plastic containers (as the name of the programme suggests) among the participants of the programme. They were distributed for free to the SEC groups 1–4, and they were sold to the SEC groups 5 and 6 and other customers. Within the next few years, the programme was expanded to cover more neighbourhoods. Between 2008 and 2014, the number of green containers distributed was 14,408: except for about 1000 replacement containers, all others were for new participants (Alcaldía Municipal de Cajicá 2015). GCP participation is not only from the households: there are other institutional participants such as industries, commercial entities, educational institutions, and government institutions (Alcaldía Municipal de Cajicá 2015). Depending on the need of the customers and the expected volume, GCP offers three container sizes: 10, 20, and 30 litres in volume. The containers incorporate a false bottom in order to collect leachates.

The frequency of the organic waste collection in Cajicá is once per week. As such, the GCP participants must mix the organic waste with a substance called bokashi – a mixture of effective microorganisms (EM) and wheat bran. When mixed with organic waste, bokashi prevents offensive odours inside the house and also catalyses the composting process (Alcaldía Municipal de Cajicá 2015). Cajicá's GCP delivers a 1.5 kg pack of bokashi to all its participants around every 2 months for free.

Cajicá Municipality was successful in letting the financial investment made towards the GCP grow continuously over time, which led to the improvement in services and the organic waste collection coverage in both urban and rural areas. The specific budget for the GCP for the period of 2016–2019 is around USD 364,000 per year (based on the average exchange rate for 2015, USD 1 = COP 2743). The budget is mainly distributed among the following activities (personal communication with Javier Rodríguez, Head of Waste Management of Cajicá's Sanitary Collection Enterprise, May 2018):

- Raw materials and workspace for bokashi manufacturing
- Cost of green containers
- Awareness-raising campaigns and programme advertising
- Trainers for awareness campaigns and capacity building (as of now GCP has 18 full-time trainers)

3.1 Organic Waste Source Separation: Awareness Raising

The success of an organic waste composting programme depends on the quality of the material received. Many waste composting programmes have failed in the past due to quality issues (Hettiarachchi et al. 2018a). One unique aspect of Cajicá's composting programme is the commitment the municipality has made to raise the awareness in organic waste source separation among all stakeholders from all sectors including households, commerce, industry, official institutions, schools, and universities. This public commitment started during the composting pilot mentioned above and has evolved in different ways over the years.

Cajicá's educational programme is now equipped with various tools and methods such as games, hands-on activities, brochures, videos, and workshops, which are useful to send an effective message about source separation of organic waste. Figure 2.5 presents one such educational poster distributed among residents to increase awareness about source separation. The poster describes in seven steps the correct procedure to handle the green containers: (1) use the false bottom container; (2) spread a spoonful of bokashi at the bottom of the container; (3) throw the separated waste on the bokashi; (4) spread a spoonful of bokashi every night; (5) compress the waste in order to release any air; (6) seal the container tightly; and (7) drain the liquids twice a week and deposit them in the siphon or use it as fertiliser for your plants after diluting with water.

Bokashi distribution mentioned above is conducted as a door-to-door service carried out by trained personnel. The GCP uses this as an in-person training opportunity to answer any question participants may have on the process. In addition, GCP has also conducted several training workshops to raise awareness across

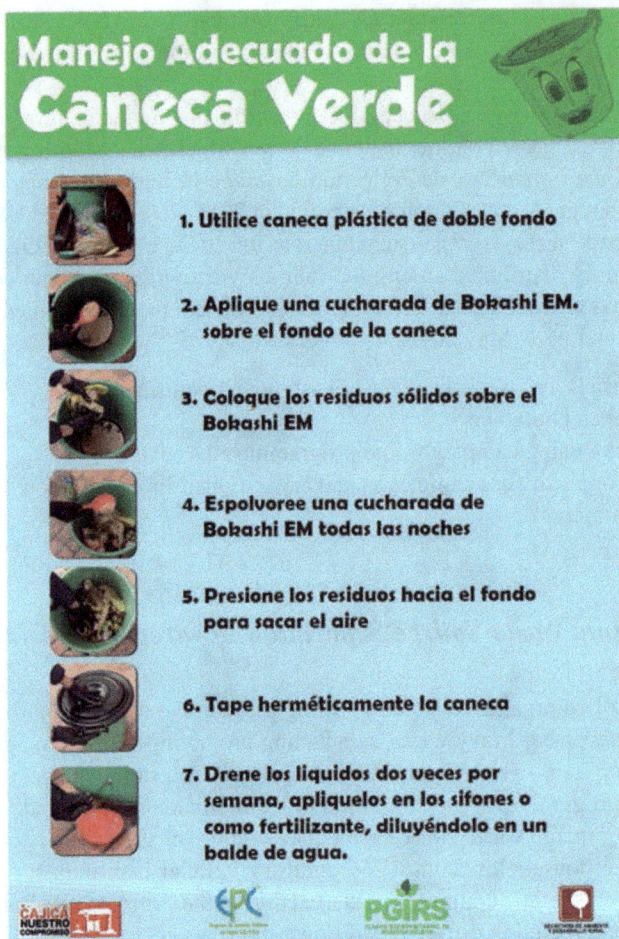

Fig. Educational brochure about source separation, Green Containers Program, and EM's good practices. (Poster courtesy: Cajicá Municipality)

various segments of society. A summary of the workshops conducted in 2017 is presented in Table. The care and attention they have put towards the next-generation participants is very much visible in the numbers associated with schools and kindergartens.

3.2   Composting Process and Quality Control

According to Cajicá's Integrated Solid Waste Management Plan (Cajicá PGIRS, hereafter), in 2015, EPC managed to increase the coverage of collection of organic waste to 95% in the urban area, although the coverage in the rural areas was still low

Table Participant breakdown for the awareness-raising workshops conducted in 2017

Customer type	Number of trained people
Residential buildings	791
Single-family households	1715
Commerce	310
Industries	383
Government institutions	112
Schools	4026
Kindergartens	1120
Total	**7177**

Source: Sanitary Collection Enterprise of Cajicá (2017)

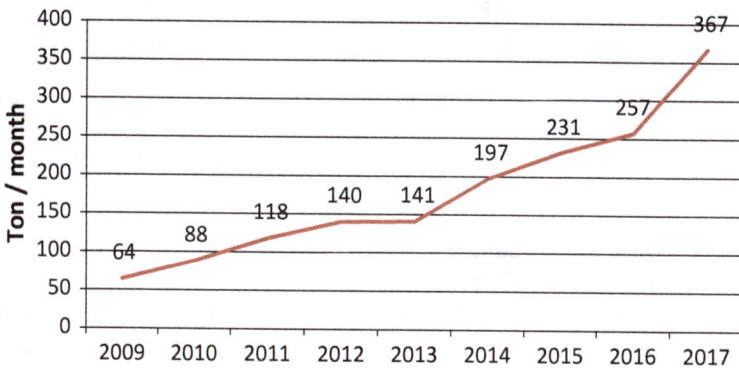

Fig. Organic waste recovery from 2009 to 2014. (Data: Sanitary Collection Enterprise of Cajicá 2017, and personal communication with EPC in Cajicá November 2018)

(Alcaldía Municipal de Cajicá 2015). The total amount of organic waste recovered is also increasing steadily as evident in Fig.. In addition to the expansion of the collection routes, there are also other reasons behind this success, such as the increased public acceptance of the programme and the awareness-raising activities (Alcaldía Municipal de Cajicá 2015).

The organic waste collected by EPC is then shipped to a facility operated by IBICOL SAS – a private vendor located in Tocancipá for composting and marketing the compost (see Fig.). The substrate consists mainly of separated organic waste from the Cajicá Municipality, agricultural waste from flower and tomato plantations, and some industrial organic waste in the form of milky sludge.

IBICOL SAS is a Colombian company in operation since 2007 and offers sustainable alternatives that support agricultural and livestock management and has a partnership with the Midwest Bio-Systems in the United States (IBICOL SAS 2018). The workflow between the Cajicá Municipality and IBICOL SAS has been established through a contractual agreement made between EPC and IBICOL

Fig. Composting in operation at the IBICOL SAS facility as organic waste received is being unloaded. After 24 h, this waste is transferred to the piles seen in the background for the next stage of processing (Photo taken by the authors in November 2018 with courtesy of the General Manager of the Facility, Mr. Francisco Pradilla)

SAS. Both parties work together to define the cost of the recovery process to be paid by EPC (Alcaldía Municipal de Cajicá 2015).

The composting workflow at the facility is divided into three parts: (1) reception zone (solid and liquid waste), (2) composting zone, and (3) compost storage zone. IBICOL SAS follows a specific set of standards and control procedures to maintain the quality of the product. They constantly measure and control temperature, humidity, aeration, and evolving CO_2. Samples from the final product are routinely subjected to laboratory testing such as elemental, proximate, and carbon-nitrogen ratio analysis for quality control purposes. The composting facility has the capacity to treat 1000 tons of organic waste each month. The current technology uses the Aeromaster PT-120 (pull-type) compost windrow turner, which works on three aspects concurrently: windrow turning and watering, addition of inoculants, and reshaping of the windrow. The average process time is around 10 weeks for the mesophyll and thermophile phases.

IBICOL SAS sells compost to the agricultural sector (mainly flower and vegetable growers in Cundinamarca); the municipalities to use in parks, green areas, and public gardens; and other sectors such as mining and landfilling. To achieve sales targets and maintain their customer base in the region, it is important that they maintain the quality of their products. IBICOL SAS conducts frequent sampling and testing to compare compost quality with the Colombian National Standards (NTC 5167). Table presents a comparison between the IBICOL SAS compost characterisation and the Colombian National Standards. It is evident from Table that except for one parameter (total oxidisable organic carbon), Cajicá's compost meets the Colombian quality standards reasonably well.

Table Typical compost characterisation and Colombian regulation standards

Parameter	Units	NTC 5167 (Colombian National Standard)	IBICOL test results
Losses due to volatilisation	%		43.5
Ash	%	≤60	22.7
Moisture	%	≤35	33.8
Total oxidisable organic carbon	%	≥15	14.5
Total phosphorus (P205)	%	Declare if greater than 1	1.09
Total potassium	%	Declare if greater than 1	1.69
C/N relation			11
Cation exchange capacity	meq/100 g	≥30	30.6
Moisture retention	%	100	120
pH		Between 4 and 9	8.25
Density	g/cm3	≤0.6	0.46
Arsenic (As)	ppm	41	ND[a]
Cadmium (Cd)	ppm	39	0.18
Chrome (Cr)	ppm	1200	7.5
Mercury (Hg)	ppm	17	ND[a]
Nickel (Ni)	ppm	420	6.13
Lead (Pb)	ppm	300	1.42

Data: Norma Técnica Colombiana (2004), IBICOL SAS (2018)
[a]*ND* no detection

4 Discussion: Lessons from Cajicá

With the information presented in the previous sections, it is fair to say that Cajicá is a case study that tells a success story. There are some interesting observations that we consider as the key reasons behind the success the project achieved in 10 years with a continuous growth. Key features are discussed briefly below and might be considered as valuable tips for other municipalities, especially in developing countries that are considering initiating composting or other types of waste recovery projects. At the end of this section, we also briefly discuss some of the potential concerns and threats that we have observed while analysing Cajicá GCP.

4.1 Proper Planning and Stakeholder Involvement

Cajicá PGIRS briefly introduced before played a vital role in shaping the waste management activities in Cajicá. The methodology for PGIRS development stresses on the need to consult other existing municipal plans (such as land use plans, basin plans, etc.) during the PGIRS design phase to achieve synergies and avoid inconsistencies (Ministry of Housing, Cities and Territory 2014). Cajicá PGIRS also

Table Organic waste management synergies between Cajicá PGIRS and the MDP from 2008 to 2019

Municipal Development Plan (MDP)	Relation with Cajicá PGIRS and Green Containers Program (GCP)
2008–2011 "Cajicá standing the change is with all"	The programme called "Expansion of coverage, reposition of networks, and improvement of quality for the provision of home public services" aimed to increase the coverage of cleaning services, implement the PGIRS, and encourage source separation and waste recovery and recycling
2012–2015 "Progress with social liability"	The Clean Municipality programme, aimed to increase the coverage of cleaning services, adjust and implement the PGIRS, increase the collection of organic waste, and implement a strong educational campaign for the GCP
2016–2019 "Cajicá, our commitment"	The programme "Efficient and quality public services for Cajicá" established the objective to strengthen the GCP, the source separation, and also the formulation and implementation of a public recovery organic waste facility

Source: Concejo Municipal de Cajicá (2007, 2011, 2015)

established this coherence between other existing plans in order to be efficient in public management and achieve a greater impact. Especially in the case of Cajicá, the other plan available is the Municipal Development Plan (MDP) made for every 4 years. Table presents a summary of organic waste management synergies we observed between Cajicá PGIRS and their MDP.

Cajicá PGIRS has improved over time. The first version of the plan was announced in 2005, and it was updated in 2010 and then updated again in 2015 due to regulation changes and new guidelines mandated by the Ministry of Housing and the Ministry of Environment. These ministerial guidelines recommend that the PGIRS design must be from an interdisciplinary perspective with participation of different stakeholders from the municipality. That was exactly what Cajicá did: all Cajicá PGIRS versions were designed with the participation of representatives from the municipality, EPC, schools, universities, public institutions, non-governmental organisations (NGOs), waste pickers, and regional environmental authority.

This experience helped Cajicá to build a stronger PGIRS and the basis for smooth implementation. For instance, in 2014 during the Cajicá PGIRS updating process, the municipality signed an agreement with CEMPRE (a Latin American NGO that works in recycling) in order to help them with the social aspects of the recycling scheme and waste picker socio-economic conditions (Alcaldía Municipal de Cajicá 2015). Likewise, some other agreements with the environmental authority (CAR) and IBICOL were developed.

For Cajicá PGIRS implementation and monitoring purposes, a team comprising of EPC and the environmental office of the municipality was established. The team followed a Plan-Do-Check-Act (PDCA) strategy to achieve better results. According to EPC, continuous practice of PDCA has helped to improve the service, reduce environmental impacts, prevent risks, and reduce costs in the processes of source separation, collection, and transport of waste (personal communication with Javier Rodríguez, Head of Waste Management of Cajicá's Sanitary Collection Enterprise, May 2018).

4.2 The Business Model and Governance Aspects

Finding the correct business model and governance structure was also crucial for the success of the GCP. A composting operation must be carried out with a business perspective and sound administrative/organisational plan. Ekelund and Nyström (2007) pointed out that it is also important to have strong working relationships with external agents who can provide technical, commercial, financial, and research support. It is important to mention that IBICOL SAS facility possesses all of the above to win a local reputation as an outstanding organic waste composting plant. Their strength as a composting facility can be attributed to the following factors: solid business model/strategy; close relationship with the municipality of Cajicá; guaranteed suppliers; high procedural and operational standards; diversification of products and customers; and partnerships with technology suppliers from the United States.

Governance structure decides how political power is exercised to manage economic and social resources for development (World Bank 1994), which is usually categorised into three types based on the type of main actors involved: bureaucratic governance, market governance, and network governance (Thompson et al. 1991). Hettiarachchi et al. (2018b) described how these three types of governance may be involved in waste management activities. In Cajicá, the GCP benefitted from the co-coexistence of mainly two types of governance: bureaucratic and network.

Bureaucratic governance (also known as hierarchy governance) is about following rules, as defined by the hierarchical authorities such as governments (Colebatch and Larmour 1993). Cajicá Municipality has achieved an elevated position within the country based on the positive bureaucratic governance it has exhibited. The Colombian Municipal Performance Measurement (MPM) is a national index that aims to measure and compare the performance of municipalities in the entire country. The index encompasses variables such as administrative efficiency, compliance with the goals of their development plans, provisioning of basic services, budget execution, improvements in the welfare of the population, transparency, and administrative and fiscal management (DNP 2016). The MPM outcome of 2016 positioned Cajicá within the 10 best municipalities out of 217 in the same category (DNP 2016). It is a tremendous achievement and an accolade for the municipality for the steps taken to guarantee improvements in the welfare of the population.

Network governance constitutes a distinct form of coordinating economic activity, which combines markets and hierarchies (Jonas et al. 1997). It is a democratic decision-making process, within the context of sustainable development and public-private partnership (PPP) (Hettiarachchi et al. 2018b). This feature was clearly visible throughout the establishment and running of the GCP. In Cajicá, stakeholders such as IBICOL SAS and EPC play active and important roles in the GCP. Their collaboration with the municipality to run the GCP has created more than 50 job opportunities for people.

PPP in Cajicá has also found to be effective in relation to the residential apartment buildings. EPC, residents' associations, and some construction companies collaborated to develop a construction guide for residential apartment buildings to ensure enhanced capacities for source separation and waste management. In the guide, the construction companies can find standards for the waste collection room, recommendations for ventilation inside the kitchens, and finally, to foster source separation, a clause that recommends against the installation of garbage chutes.

4.3   Educating the Next Generation

Environmental education among the public has played an important role in the process of development and maintenance of GCP. Cajicá has implemented different strategies for environmental and waste management education, especially through programmes launched at schools and universities. The basis of the successful source separation model of organic waste was explained to the public through the Decree 061 in 2005 (called "Citizen culture for the integral management of solid waste and the program of separation at the source") and then further enhanced through the Decree 003 in 2011.

As seen in Table, the municipality has a strong commitment towards awareness raising among school children. One of the aims of this effort is to enable children to take the "organic waste recovery message" home to educate their families. The GCP approaches students through some composting and farming projects at schools and catches their attention with the learning-by-doing concept. It has provided conclusive evidence that such education programmes can play a key role in developing children's knowledge about sustainable waste management. The project also concludes that if the message delivered in school is taken home, waste management at homes becomes more sustainable. In addition, EPC organises "Eco-Arte" an annual school contest on the topic of waste management through arts in different categories. The idea is to construct artwork with recycled materials during the school year, and then in November, the best school wins a prize. This practice has helped them make source separation more popular and improve environmental awareness.

One of the most innovative aspects of the education programme in Cajicá is the work done by the trainers. There are about 18 trainers who are specialised in addressing different customer groups (residential, commerce, industry, public, etc.). It is important to note that they have appointed such specific trainers for schools and universities, too. There is also one trainer working exclusively on social media and advertising, who is able to catch the attention of the younger generation, especially students.

4.4 Technology Adaptation to Local Conditions

Technology transfer from developed and developing countries does not always go as planned due to the complexity of the technology or the lack of capacity building at the final user. On the other hand, correctly managed technical adaptation can increase the amount of materials recovered, reduce problems in operation and maintenance of equipment and machinery, and positively influence systems sustainability (Oviedo-Ocaña et al. 2016). In this context, adaptation of technology to the local settings was a key part of the success of Cajicá. One might argue that composting is a very well-established process that does not need much adaptation. This is true about the compost making process. However, some fine-tuning is still needed to arrange properly all the parts of the process, regarding the logistical affairs as well as source separation and collection.

In this case, we found two such successful adaptations: one is the green container design, and the other is the bokashi mix. The green containers were designed by gauging the average amounts of waste generated by households and other places. The container also includes a false bottom to contain leachate and a hermetic lid to prevent odours. This simple but effective design helped GCP to receive the much-needed public acceptance and improve source separation and collection volumes.

In addition, bokashi used in the containers is also another example for such technology adaptation. EPC conducted a number of trial runs to find the correct combination of bokashi needed. What they use now is the best they found to prevent offensive odours, retain high amounts of leachate, and prevent excess fungi and fruit fly growth (personal communication with Javier Rodríguez, Head of Waste Management of Cajicá's Sanitary Collection Enterprise, May 2018).

4.5 Putting Nexus Thinking into Practice

Nexus thinking is a concept that has become popular recently within the environmental resource management circles. It is a way to improve the security of resources by integrating management and governance across sectors and scales, thus reducing trade-offs and building synergies with the overall intention of promoting sustainability (Hoff 2011). Any composting project is a good example to explain the benefits of nexus thinking (Sallwey et al. 2017). When GCP was introduced years ago, Cajicá Municipality may not have considered any aspect of nexus thinking, as the concept itself was in its infancy back then. However, looking back now at how the GCP project had been designed and developed, it is clear that they have employed nexus thinking to address both resources and governance aspects of the project.

From the material resources point of view, cross-sectoral integrated management benefits from synergies. In other words, sustainable management of one resource can also alleviate resource issues in another sector. In this case, the organic waste is a resource that originated from the waste sector, and the compost made from it provides nutrients to the agricultural sector. The nexus application in

Fig. The organic farm located next to IBICOL facility uses compost produced by IBICOL SAS. This is a good example explaining how the agricultural sector benefits from organic waste recycling. (Photo taken by the authors in November 2018 with courtesy of the Facility Manager, Mr. Andrés Botero)

Cajicá also explains the economics of it through cost savings. The municipality is saving money through the reduced amounts of waste that is now going to the land-fill. Saving landfill volume is directly visible, but there are other cost savings that might not be easily visible, through transport, oil, drivers, vehicle maintenance, transit tolls, change of tyres, etc. (personal communication with Javier Rodríguez, Head of Waste Management of Cajicá's Sanitary Collection Enterprise, May 2018).

The other aspect of nexus thinking is the policy integration through active stake-holder involvement. As explained in the previous sections, the municipality con-sulted all other existing policies and involved all stakeholders to come up with the new plan for the GCP development.

4.6   Concerns

Built on the experience that spans over a decade, GCP runs smoothly now. Despite all the achievements and strengths of the Cajicá case as described above, there are some important concerns to tackle. In a recent publication, Hettiarachchi et al. (2018a) presented some key limitations and potential issues GCP might face.

They discussed particularly about the financial wellbeing of the GCP and its high dependency on the municipality. Currently, the organic waste collection system is supported by the municipal government – both financially and politically. As of now the municipality spends about USD 364,000/year to cover the cost of the programme, creating a huge risk in terms of maintaining its stability. This means that any financial issues faced by the municipality can put the programme in jeopardy. Also, this municipal support is subjected to the changes in political will and priorities. The political leadership of the municipality changes every 4 years, resulting in new staff and agenda. Fortunately, mayors in the past 10 years recognised GCP as a priority, but if any new mayor in the future does not see it the same way, he/she can stop the programme or restrict financial assistance. Hettiarachchi et al. (2018a) also suggested the need to introduce a policy instrument where a nominal surcharge is added to their waste collection or utility bill as this works better in the region, to raise money to help the programme achieve financial freedom. Part of the money can be used to reward customers through food vouchers.

Hettiarachchi et al. (2018a) also discussed the rapid population growth in Cajicá and the implications of the resulting increase in waste volume on the GCP. As for the projections made for 2018, the population has increased by 33% since 2005, and based on Cajicá's household classification statistics (DANE 2014), the middle-class population is growing rapidly. This has resulted in a rapid increase in the waste generation per capita. One person in Cajicá produced 0.58 kg of waste per day on average in 2009, which has increased to 0.84 kg by 2014 (Alcaldía Municipal de Cajicá 2015). The rapidly increasing population and waste volume may not allow municipality staff to continue to visit every household door to door to collect organic waste and educate people. The current system already requires a lot of manpower, which will become more challenging to maintain when the municipality must serve a much larger population in the near future. Training citizens to bring their source-separated organic waste to centres established by the municipality may relieve the programme from some of the financial obligations (Hettiarachchi et al. 2018a).

The excessive dependency GCP has on external parties is a concern known to EPC (based on personal communication with Cajicá's Sanitary Collection Enterprise, May 2018). They have realised the risk of failure, in the case of composting vendor (IBICOL SAS) failing to do their part in the programme due to reasons such as technical issues, declaring bankruptcy, etc. EPC or the municipality currently do not have a contingency plan for such an event. In the worst-case scenario, the only action they could take is to send the organic waste to a landfill.

5   Summary and Conclusions

In this chapter, we presented, reviewed, and critiqued Cajicá's experience in recovery of organic waste to make compost. The key aspects of their success as well as some of the potential concerns were also identified. A summary of our main observations is presented below:

- Cajicá's Green Containers Program (GCP) was carefully designed with the tools, personnel, and economic resources made available by the municipality. A strong and multidisciplinary planning process was used to achieve the intended goals.
- Coordination between the stakeholders such as the municipality, Sanitary Collection Enterprise (EPC), schools/universities, public institutions, NGOs, waste pickers, and regional environmental authorities was fundamental to collect different viewpoints and advice, which ultimately strengthened the foundation of the project.
- The decision to first run a composting pilot project was commendable. It helped Cajicá to gain some first-hand experience before launching the long-term project.
- Awareness raising in source separation of waste and composting process was conducted using multiple channels. This helped the programme tremendously by educating the public and positively influencing their behaviour towards proper waste management and winning public acceptance. The efforts made to educate the next generation through involving schools and universities is certainly an investment for the future.
- Cajicá's case study also teaches us a lesson about the correct choice of business model. They decided to opt for a network governance model to create a private-public partnership (PPP).
- Technology transfer has been a key challenge for waste management projects launched in developing countries. In this context, another lesson we can learn from Cajicá's experience is how well they adapted technology to the local settings.
- Undoubtedly, Cajicá's GCP is a good example to explain the benefits of nexus thinking, where integrated management helps both the waste and agricultural sectors to prosper.
- High dependency of GCP on the municipality for financial support and the dependency on a single external partner for the composting process could be viewed as some concerns that may raise issues in the future.

References

Adhikari, B. K., Barrington, S., Martínez, J., & King, S. (2009). Effectiveness of three bulking agents for food waste composting. *Waste Management, 29*, 197–203.

Alcaldía de Barranquilla. (2015). *Plan de Gestión Integral de Residuos Sólidos –PGIRS 2016-2027.* Barranquilla, Colombia.

Alcaldía de Bogotá. (2015). *Plan de Gestión Integral de Residuos Sólidos –PGIRS 2016–2027.* Bogotá, Colombia.

Alcaldía de Medellín. (2015). *Plan de Gestión Integral de Residuos Sólidos –PGIRS 2016–2027.* Medellin, Colombia.

Alcaldía de Santiago de Cali. (2015). *Plan de Gestión Integral de Residuos Sólidos –PGIRS 2016–2027.* Santiago de Cali, Colombia.

Alcaldía Municipal de Cajicá. (2015). *Actualización del Plan de Gestión Integral de Residuos Sólidos 2016–2027.* Cajicá: Municipio de Cajicá.

Alcaldía Municipal de Cajicá. (2018). *Municipio de Cajicá.* Cajicá, Colombia. www.cajica.gov.co/informacion-general. Accessed 8 May 2018.

Banco Mundial. (2015). *Estrategia Nacional de Infraestructura – Sector Residuos Sólidos*. Bogotá: Consultancy, Departamento Nacional de Planeación.

Colebatch, H. K., & Larmour, P. (1993). *Market, bureaucracy and community: A student's guide to organization*. London: Pluto Press. ISBN 0745307620.

Concejo Municipal de Cajicá. (2007). *Plan de Desarrollo Municipal 2008–2011*. Cajicá: Municipio de Cajicá.

Concejo Municipal de Cajicá. (2011). *Plan de Desarrollo Municipal 2012–2015*. Cajicá: Municipio de Cajicá.

Concejo Municipal de Cajicá. (2015). *Plan de Desarrollo Municipal 2016–2019*. Cajicá: Municipio de Cajicá.

DANE. (2005). *Boletín Censo general*. Bogotá: Departamento Administrativo Nacional de Estadística.

DANE. (2013). *Encuesta de Calidad de Vida – Presentación general*. Bogotá: Departamento Administrativo Nacional de Estadística.

DANE. (2014). *Encuesta Multipropósito – Boletín Técnico*. Bogotá: Departamento Administrativo Nacional de Estadística.

DNP. (2016). *Nueva Medición de Desempeño Municipal – Resultados*. Departamento Nacional de Planeación. colaboracion.dnp.gov.co/CDT/Desarrollo%20Territorial/MDM/Resultados%20 Desempe%C3%B1o%20Integral%202016.xlsx?Web=1. Accessed 17 May 2018.

Ekelund, L., & Nyström, K. (2007). *Composting of municipal waste in South Africa – Sustainability aspects*. Uppsala: Uppsala Universitet. ISSN: 1650-8319, UPTEC STS06 012.

Hettiarachchi, H., Meegoda, J. N., & Ryu, S. (2018a). Organic waste buyback as a viable method to enhance sustainable municipal solid waste management in developing countries. *International Journal of Environmental Research and Public Health, 15*(11), 2483. https://doi.org/10.3390/ ijerph15112483.

Hettiarachchi, H., Ryu, S., Caucci, S., & Silva, R. (2018b). Municipal solid waste management in Latin America and the Caribbean: Issues and potential solutions from the governance perspective. *Recycling, 3*(2), 19. https://doi.org/10.3390/recycling3020019.

Hoff, H. (2011). *Understanding the Nexus, background paper for the Bonn 2011 conference*. Stockholm: Stockholm Environment Institute.

IBICOL SAS. (2018). https://www.ibicol.com.co/insumos. Accessed 14 May 2018.

Jonas, C., Hesterley, W., & Borgatti, S. (1997). A general theory of network governance: Exchange conditions and social mechanisms. *Academy of Management Review, 22*, 914.

Kaza, S., Yao, L., Bhada-Tata, P., & Van Woerden, F. (2018). *What a waste 2.0: A global snapshot of solid waste management to 2050* (Urban Development Series). Washington, DC: World Bank.

Ministry of Housing, Cities and Territories. (2014). Residuos sólidos/metodología PGIRS. *Minvivienda*. www.minvivienda.gov.co/Residuos%20Solidos/ Metodolog%C3%ADa%20PGIRS.pdf. Accessed 15 May 2018.

Ministry of Housing, Cities and Territories. (2015). Decree 1077. *Diario Oficial de Colombia*. Bogotá, Colombia.

Norma Técnica Colombiana. (2004). *Productos para la industria agrícola productos orgánicos usados como abonos o fertilizantes y enmiendas de suelo*. ICONTEC NTC 5167, 15 de June de 2004, Bogotá, Colombia.

OECD/UN ECLAC. (2014). *OECD environmental performance reviews: Colombia 2014*. Paris: OECD Environmental Performance Reviews, OECD Publishing. https://doi.org/10.178 7/9789264208292-en.

Ospina, D. (2009). *Aprovechamiento Y Valorización De Residuos En Cuatro Municipios De Cundinamarca*. Postgraduate Thesis, Bucaramanga: Universidad Industrial de Santander.

Oviedo-Ocaña, E. R., Dominguez, I., Torres-Lozada, P., Marmolejo-Rebellón, L. F., Komilis, D., & Sanchez, A. (2016). A qualitative model to evaluate biowaste composting management systems using causal diagrams: a case study in Colombia. *Journal of Cleaner Production, 133*, 201–211.

Pace, S. A., Yazdani, R., Kendall, A., & Simmons, C. W. (2018). Impact of organic waste composition on life cycle energy production, global warming and water use for treatment by anaerobic digestion followed by composting. *Resources, Conservation & Recycling, 137*, 126–135.

Red de ciudades cómo vamos. (2014). *Informe sobre la política pública de inclusión de recicla-dores de oficio en la cadena de reciclaje: Informe regional*. Bogotá: Fundación Avina.

Sallwey, J., Hettiarachchi, H., & Hülsmann, H. (2017). Challenges and opportunities in munici-pal solid waste management in Mozambique: A review in the light of Nexus thinking. *AIMS Environmental Science – Special Issue: Waste recycling, Reduction and Management, 4*, 621–639. https://doi.org/10.3934/environsci.2017.4.621.

Sanitary Collection Enterprise of Cajicá. (2017). *Waste management report submitted to the municipal council of Cajicá*. Cajicá: Sanitary collection Enterprise.

Superintendence of Public Services. (2017). *National report of solid waste final disposal in 2016: National report*. Bogotá: Superintendence of Public Services.

Thompson, G., Frances, J., Levacic, R., & Mitchell, J. (1991). *Markets, Hierarchies and Networks*. London: Sage.

UN Environment. (2017). *Solid approach to waste: How 5 cities are beating pollution*. United Nations Environment Program Website. www.unenvironment.org/news-and-stories/story/solid-approach-waste-how-5-cities-are-beating-pollution. Accessed 8 May 2018.

USEPA. (2016). *Inventory of greenhouse gas emissions and sinks: 1990–2014*. United States Environmental Protection Agency Website. 15 de April de 2016. 19january2017snapshot. epa.gov/sites/production/files/2016-04/documents/us-ghg-inventory-2016-main-text.pdf. Accessed 26 June 2018.

Wei, Y., Li, J., Shi, D., Liu, G., Zhao, Y., & Shimaoka, T. (2017). Environmental challenges imped-ing the composting of biodegradable municipal solid waste: A critical review. *Resources, Conservation and Recycling, 122*, 51–65.

World Bank. (1994). *Governance, the World's Bank experience*. World's Bank activities, Washington, DC: World Bank.

6

Composting in Sri Lanka: Policies, Practices, Challenges, and Emerging Concerns

Warshi S. Dandeniya and Serena Caucci

Abstract Compost is a widely accepted organic fertiliser throughout the world. It is being produced using a wide variety of source materials at household to commercial scale. With the increased population and changes in food consumption pattern tending towards a vegetable- and meat-rich diet, the amount of organic waste generated in urban and peri-urban settings has increased. Many governments promote composting as a process that helps them to reduce the volume of organic waste and recycle nutrients back to croplands. Some examples of organic waste accumulated in large scale include household waste from urban and peri-urban settings, sewage, animal farm waste, agricultural waste from large-scale markets, food debris, and kitchen waste from hotels. The composition of compost varies in a wide range depending on the nature of materials used to produce it. The safety concerns related to compost also vary along the same line. The quality of compost has become a subjective term that means different aspects to different bodies due to a lack of commonly agreed standards to regulate the composting process and the final product itself. Recent research findings indicate that compost can serve as a carrier of potentially toxic trace elements, organic pollutants, and determinants of antimicrobial resistance to the environment and along the food chain. Producing good-quality compost safe to human health and the environment at large has become a challenge that should be addressed at various levels: from production to policymaking. This chapter discusses some of the major challenges faced in Sri Lanka with compost making. To prepare the background for this discussion, information on the policies and current practices of nutrient management in Sri Lanka is also presented. The context may be applicable to many other developing countries in the tropics.

W. S. Dandeniya (✉)
Department of Soil Science, Faculty of Agriculture, University of Peradeniya, Peradeniya, Sri Lanka
e-mail: warshisd@pdn.ac.lk

S. Caucci
United Nations University (UNU-FLORES), Dresden, Sachsen, Germany
e-mail: caucci@unu.edu

Keywords Animal manure · Antibiotic resistance · Compost · Fertiliser ·
Municipal solid waste · Organic pollutants

1 Introduction

Compost can be used as a nutrient source in crop cultivation and often increase
productivity when used in combination with chemical fertilisers. It is a popular
fertiliser in organic farming and in integrated nutrient management systems.
However, the long-term use of compost in large quantities and/or application of
poor-quality compost to soil can deteriorate environmental quality and pose a threat
to the safety of food (Deportes et al. 1995; Garcia-Gil et al. 2000; Smith 2009,
2018). The progressive accumulation of toxic trace elements such as Pb and Cd in
soils has been reported in several studies where long-term application of compost
produced from municipal solid waste (MSW) was applied (Garcia-Gil et al. 2000;
Smith 2009). The contamination of food items with potentially toxic trace elements
and human pathogens due to the application of compost to crops has been reported
in literature (Deportes et al. 1995; Johannessen et al. 2004; Smith 2009; Maffei
et al. 2013).

Several types of composts are being produced and these differ in terms of starting material combinations and methods to cater to unique requirements. For example, easily biodegradable and nutrient-rich wastes like animal manures are used to
produce compost, when compost is required as a source of nutrients to substitute
chemical fertilisers (Garcia-Gil et al. 2000). Biochar like more recalcitrant carbon
has been added to compost lately, when applied to cultivate soil fertility in the long
term (Agegnehu et al. 2015). Municipal solid waste composting is commonly
known as a method for reducing and recycling waste, and the product is being used
in landfills and for environmental restoration purposes (Hurst et al. 2005). Therefore,
the quality and safety concerns of compost heavily depend on the nature of materials used to produce compost as well as on the intended goals of end use (Garcia-Gil
et al. 2000; Hurst et al. 2005; Smith 2009; Agegnehu et al. 2015). The use of compost such as those made using MSW, with a risk of containing potentially toxic
contaminants, is often encouraged to be limited to contained environments or in
systems that have no direct link to the human food chain. However, there is no consistency in how people use those products as some use such compost in landfills,
while some apply it in crop fields. In many countries, there is only one set of standards regarding the composition of compost and common guidelines to follow during marketing irrespective of the diversity in the nature of materials used, method of
composting, and intended end use. Further, the standards are limited to well-known
parameters like moisture and sand contents; contents of nutrients as N, P, and K; and
a few potentially toxic trace elements and carbon content. There are no standards for
organic contaminants including volatile compounds and antibiotics.

Compost is comprised of a mixture of complex forms of nutrients and microorganisms, and once applied to soil, it replenishes the soil nutrient pool while

improving several soil fertility aspects. Therefore, compost application is highly recommended to soils with declined fertility status to improve productivity levels. Soils occupying the agricultural lands in Sri Lanka are inherently low in organic matter content and overall soil fertility. Compost as an organic fertiliser could be used to improve both the organic carbon pool and crop productivity in these soils in Sri Lanka. Understanding the gaps in knowledge on the safety and quality aspects of the composting process and final product itself is very important to promote responsible production and use of the material. In this context, this chapter discusses the policy aspects, the current practices with an emphasis on quality and safety concerns, and the challenges of compost making in Sri Lanka.

2 Agricultural Nutrient Management Practices in Sri Lanka

Historical evidences indicate that traditionally Sri Lankan farmers used organic inputs such as animal manures, crop residues, and green manures and biodynamic farming techniques, crop rotations, shifting cultivation, and crop diversification as measures for managing soil fertility. Colonisation by the Portuguese and Dutch followed by the British from the mid-sixteenth to mid-nineteenth century gradually changed the land-use pattern in the country, mainly transforming traditional subsistence farming systems to export-based commercial plantation agriculture (Mapa 2003). Organic and biodynamic means of farming continued for locally consumed food crop production until the introduction of high-yielding crop varieties and synthetic fertilisers for nutrient management with the 'green revolution' in the 1960s, aiming to improve crop yields to meet the growing demand for food (Nagarajah 1986; Palm and Sandell 1989). Some farmers used an integrated approach in nutrient management by combining biodynamic methods and organic farming techniques with synthetic fertiliser application. Later research entities and the Department of Agriculture identified the importance of finding integrated soil fertility management approaches for nutrient management due to growing concerns about nutrient depletion and overall fertility decline in agricultural soils managed with synthetic fertiliser inputs. With advancements in knowledge on soil nutrient dynamics and the interactive role of soil physical, chemical, and biological properties on plant performances, a paradigm shift in nutrient management took place across the globe in the 1980s from external input management to integrated soil fertility management (Sanchez 1997). As a result, the composting techniques, compost application rates, and benefits of applying compost were researched extensively in many parts of the world, including in Sri Lanka. In the late 1990s, compost application rates were incorporated to fertiliser recommendations of some food crops in Sri Lanka. However, organic fertiliser application was not popular among farmers due to a lack of availability of materials, poor motivation to produce compost on-farm, and issues with timing of compost production with material and resource availability for production and demand for application.

Recommendations of NPK fertilisers in Sri Lanka for different crops range from 125 to 1500 kg/ha per season (for legumes and potato, respectively, with recommendations for other annual crops and plantation crops varying within this range) (Weerasinghe 2017). However, the usage of synthetic fertilisers in the country varies between 80 and 700 kg/ha (as total of urea + triple super phosphate + muriate of potash) depending on the crop sector, with a national average of about 270 kg/ha. Despite various programmes the government implements, fertilisers are not equally available for all farmers. A fertiliser subsidy is given to rice farmers, yet a high variability in fertiliser usage per hectare basis can be observed among farms. Therefore, while some farmers apply adequate levels or more than the recommended levels of fertilisers for some crops, others are not applying adequate inputs to replenish the soil nutrient pool (Kendaragama et al. 2001; Kendaragama 2006). A number of socioeconomic constraints and poor agricultural extension services are causing these discrepancies. These facts reiterate the need for improving on-farm nutrient management using an integrated nutrient management approach (combining synthetic and organic fertilisers). In most of the annual cropping systems managed with synthetic fertiliser inputs, the application of 10 Mg of organic fertiliser per hectare has been recommended (Wijewardena 2005). Compost application has been recommended for plantation crops such as rubber and coconut as well as tea (TRI 2000, 2016; Samarappuli 2001; Tennakoon and Bandara 2003).

Poor management of synthetic fertilisers in crop production systems in Sri Lanka caused a number of environmental and health issues, including contamination of surface and groundwater resources. Although highly debated yet, many allegations suggest agricultural chemicals to be the cause of increased cancer incidents and chronic kidney disease with unknown aetiology in the country. On these grounds, environmentalists campaigned for reducing the usage of fertilisers and other synthetic chemicals in the country. About 92% of the synthetic fertiliser quantity used in Sri Lanka is imported into the country. Nearly 70% of the imported fertilisers are used in rice cultivation, and this quantity is heavily subsidised (Weerahewa et al. 2010; Wijetunga and Saito 2017). Thus, the direct cost on importation of synthetic fertilisers and the virtual cost for recovering from environmental damage caused by a mismanagement of fertilisers increase the burden on the government. The development policy of the government between 2010 and 2015 targeted to reduce the import of synthetic fertilisers by 15% by 2015 by promoting the use of organic fertilisers like compost for cropping. However, this target was not achieved. The development policy of the current government is also aiming to reduce the import of synthetic fertilisers. Government policies on the fertiliser subsidy scheme strongly influence the quantities of fertilisers imported into Sri Lanka (Weerahewa et al. 2010; Wijetunga and Saito 2017). Fertiliser imports into Sri Lanka reduced from 816,900 Mg in 2015 to nearly 548,100 Mg in 2017, which is about a 33% drop. This happened in parallel to a government policy change, which is the conversion of material subsidy given to rice farmers to a cash subsidy, encouraging the use of organic and soil test-based fertilisers. However, whether the target yields are achieved with reduced fertiliser distribution and how effective the policy changes have been are yet to be analysed.

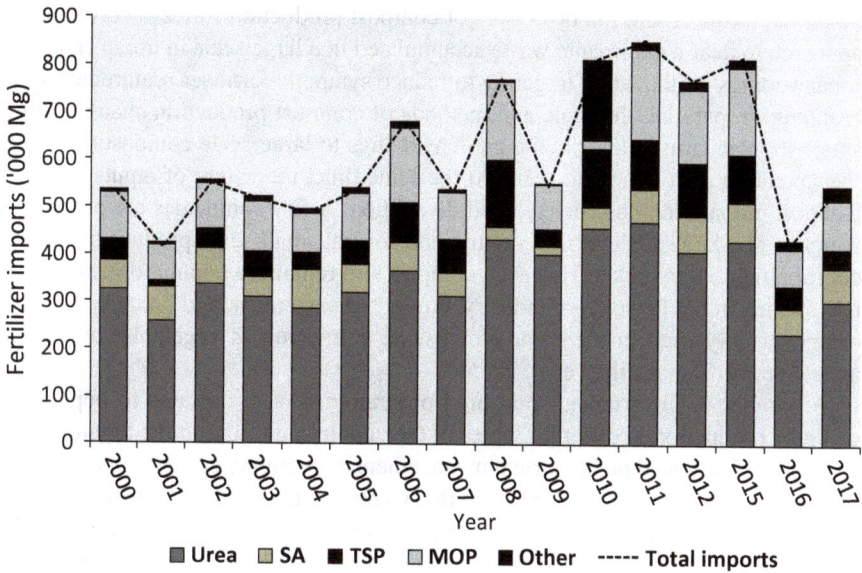

Fig. Fertiliser imports into Sri Lanka. The fertilisers include urea, sulphate of ammonia (SA), triple super phosphate (TSP), muriate of potash (MOP), and other types (zinc sulphate, kieserite, etc.). (Source: Personal communication with National Fertilizer Secretariat in 2017)

3 Government Initiatives to Promote Composting

Composting in Sri Lanka first started as a method to replace the use of synthetic fertilisers with organic fertilisers in small-scale farms or at the household level. Composting provided an opportunity to recycle crop residues and farm waste, enabling farmers to recycle nutrients at the farm scale. The input materials usually composed of crop residues, cattle manure, wood ash, grass clippings, some soil or compost (as an inoculant of microorganisms), and rice straw like crop residues. Despite the intensive campaigns made by the Department of Agriculture in Sri Lanka to popularise compost production, it still remains less popular among farmers. Seasonality and spatial heterogeneity in the distribution of materials required for composting (crop residues) and labour requirement are among the reasons for the weak adoption of composting technology. In addition, indirect facts such as the high availability of chemical fertilisers on the market, the quick response the crops exhibit for chemical fertiliser applications, the requirement for compost in large quantities to improve productivity under tropical climate, and a lack of awareness about the benefits of compost for soil fertility also contribute to a weak interest in composting.

Governments since the late 1990s started to promote compost production. Therefore, the objectives for promoting compost production evolved with time: first as a method for providing an organic fertiliser source to farmers and later as a way to provide a solution for recycling organic waste generated at household level and in animal farms. Then, the objectives of compost production further evolved as an approach to deal with organic waste accumulated in a large scale in urban and peri-urban settings all the while targeting to reduce synthetic fertiliser requirements for cropping. In parallel, the scale and methods of compost production changed from small-scale at farm-field or home garden settings to large-scale composting facilities operating at commercial scale. At the same time, the nature of inputs used for compost preparation has changed and diversified. A few companies are producing compost at a large scale in the country, and several small-scale producers are also contributing to the market. However, compost still remains a popular organic fertiliser applied in the floriculture industry, nursery management, and home gardening, but not widely used in the commercial-scale cultivation of vegetables and other annual crops (Samarasinha et al. 2015).

A number of government-sponsored programmes were initiated to popularise compost production at different scales in the country and to promote organic agriculture and integrated plant nutrient management. Government programmes such as 'Pilisaru' implemented in 2008 by the Central Environmental Authority and Ministry of Environment and Renewable Energy of Sri Lanka targeted to recycle biodegradable waste at the domestic and municipal scale (Basnayake and Visvanathan 2014). The Pilisaru programme also promoted the sorting of waste into different categories as biodegradable, paper, plastic/polythene, glass, and metal before collection. Through this programme, nearly 18,000 compost bins were distributed freely among people in urban and peri-urban settings mainly, aiming to reduce the loading of waste to local authorities. Non-governmental organisations (NGOs) also distributed compost bins among households. During the same period, several private companies (e.g. Wayamba Polymers, CIC Pvt. Ltd., ARPICO Plastic Ltd.) started selling compost bins. By 2011 nearly 70,000 bins were distributed or sold to the citizens. While promoting compost production at domestic level, the Pilisaru programme also supported compost making with solid waste and large-scale composting by private companies/entrepreneurs. Accordingly, 100 composting stations associated with municipal councils were established by 2011. Further, the national policy and strategy on solid waste management (SWM) in Sri Lanka were developed in 2007 and 2008, respectively. The national strategy aimed to reduce, reuse, and recycle solid waste. The reduction of waste generation was achieved only in some municipal councils (e.g. Balangoda) in the country (Samarasinha et al. 2015).

The 'Api wawamu rata nagamu' (i.e. let's grow and develop the country) programme later promoted by the Department of Agriculture of Sri Lanka as part of the development policy of Sri Lanka from 2008 to 2015 promoted home gardening using the concepts of organic agriculture and integrated plant nutrient management. This programme increased the usage of compost among people. However, the overall goal of reducing fertiliser imports into the country by promoting organic agriculture and home gardening was not achieved due to several reasons. Although the Pilisaru project promoted the establishment of small- to medium-scale compost production centres in association with solid waste collecting units of municipal councils, the marketing structure for the produce was not clearly defined. Therefore, composting MSW has become a burden rather than an income-generating activity (Samarasinha et al. 2015). Further, lack of sorting waste before disposal increased labour requirement for processing compost as well as reduced the quality of compost (Samarasinha et al. 2015). Due to bad odour of decomposing waste, mixing of leachate coming out during composting to surface water and groundwater, and increased density of wild animals (e.g. stray dogs, crows, etc.) and flies, the composting centres were not well-received by people in some regions (Premachandra 2006; Samarasinha et al. 2015). Protests against composting centres were common when they were located in densely populated areas. As a result, people were not motivated to develop composting as a business in many instances, and the compost produced was not effectively used.

The policy documents of the government after 2015 also advocate for the use of organic fertilisers. State patronage was proposed to be provided to produce pest control liquids or powder utilising indigenous herbal extracts, and initiate measures to formulate a time frame for the gradual elimination of the use of chemical fertilisers and agro-chemicals. In the government's 8-year economic development plan 'Vision 2025 – A Country Enriched' (2017), the forethought for developing national policy on food quality and the use of permitted levels of fertilisers, increasing the share of organic products in the market, and improving solid waste management strategies were stated. Providing a material subsidy (synthetic fertilisers) discouraged farmers to use organic fertilisers (Weerahewa et al. 2010; Wijetunga and Saito 2017); thus, in 2017, a cash subsidy was given with the hope that this would increase the practice of organic agriculture and use of compost. Although this resulted in a clear decline in the importation and usage of synthetic fertilisers, the target of promoting compost application was not satisfied. The government policies shifted from a cash subsidy scheme to a material subsidy scheme again in late 2018. The highly dynamic socio-political environment of the country is not supportive of implementing the decisions and achieving the visions stated in the policy documents.

3.1 Standardisation and Recent Changes

Since 2017 a strong drive is evident in the country to sort garbage prior to disposal, and that has seen success in many places. Sorting garbage at the domestic level helps to reduce labour requirement and improve efficiency of MSW compost preparation (Samarasinha et al. 2015). With the increased availability of compost produced by different entities at different scales in the market, the need for controlling quality was highlighted. The standards for compost produced from MSW and agricultural waste were introduced in 2003 (SLSI 1246: 2003) by the Sri Lanka Standards Institution . Packaging and marking standards were also established. However, there is a great need for revisiting the set of standards developed to ensure the quality and safety of products because the nature of materials used for

Table   Standards for compost produced from MSW and agricultural waste

Parameter	Threshold values
Physical requirement	
Moisture (%)	<25 (dry weight)
Colour	Brown/grey to dark black
Keeping quality	Should be able to store in prescribed package condition under room temperature for 12 months
Odour	Shall not have any unpleasant odour
Sand content (%)	<10 (dry weight)
Residue (>4 mm particles) (%)	<2 (dry weight)
Chemical requirement	
pH	6.5–8.5
Total nutrient contents (%)	
Organic carbon	>20.0
Nitrogen	>1.0
Phosphorus as P_2O_5	>0.5
Potassium as K_2O	>1.0
Magnesium as MgO	>0.5
Calcium as CaO	>0.7
C:N ratio	10–25
Heavy metals (ppm)	
Cd	<10
Cr	<1000
Cu	<400
Pb	<250
Hg	<02
Ni	<100
Zn	<1000
Biological requirement	
Viable weed seeds	<16 viable weed seeds/m^2
Microbiological requirement	
Faecal coliforms per g	Free
Salmonella per 25 g	Free

Source: Sri Lanka Standard Institution (2003)

composting and the scale of operations diversify to a great length since its introduction . Especially composting unsorted MSW, which consists of a wide variety of materials, introduces a number of concerns regarding the quality and safety of the compost produced as discussed later in this chapter. Accordingly, the quality standards are being revised, and a draft document was made available for the public to comment in August 2017.

Table Changes in commonly used material for composting and scale of operation with time

Period	Scale(s) of operation[a]	Commonly used materials	Methods of composting
Before 1990s	Small scale at domestic level	Household and garden waste	Small pits/heaps, bins
1990 to 2007	Small scale at domestic level	Household and garden waste, agricultural waste (crop residues, cattle manure)	Bins, small heaps/ windrows
	Small scale by agri-businesses		
After 2007	Small scale at domestic level	Household and garden waste, agro-industry waste (crop residues, cattle manure, poultry manure), ash from dendro-power plants, MSW from urban and peri-urban settings, night soil	Domestic scale: bins, small heaps
	Small and medium scale by agri-businesses, some municipal councils, and a few companies		Small to medium commercial scale: windrows, semi-aerobic trench, in-vessel composting (e.g. inclined step-grate composter)

[a]The scale of operation of the business is defined by the capacity of handling organic waste per day: small scale, less than 10 MT/day; medium scale, 10–50 MT/day

3.2 Stakeholder Involvement: Good Practice Examples

Establishing composting as a profitable business can be achieved only if all stakeholder groups are involved in developing the process (Sinnathamby et al. 2016). Composting centres at Balangoda and Weligama municipalities are examples of successfully operating facilities that produce compost using MSW in Sri Lanka (Samarasinha et al. 2015).

Balangoda Composting Facility

This centre was established with the financial support of the World Bank, Provincial Government, Chief Minister of the provincial council, and the Central Environmental Authority of Sri Lanka (under *Pilisaru* project). Some key traits that made the Balangoda composting facility a successful operation are indicated here:

- Using waste that is in the early stage of decomposition – waste collection is being practised on a daily basis in the morning hours
- Involving key stakeholders in the process – local municipal council involved schools in the region and conducted special programmes with the involvement of key stakeholders in the region generating solid waste to promote reducing and recycling waste. As a result, the amount of waste generated reduced remarkably, and people practise sorting of waste as per the instructions which helped to reduce labour requirement for compost production
- Taxing people or entities generating mixed waste
- Not collecting hospital waste for compost making

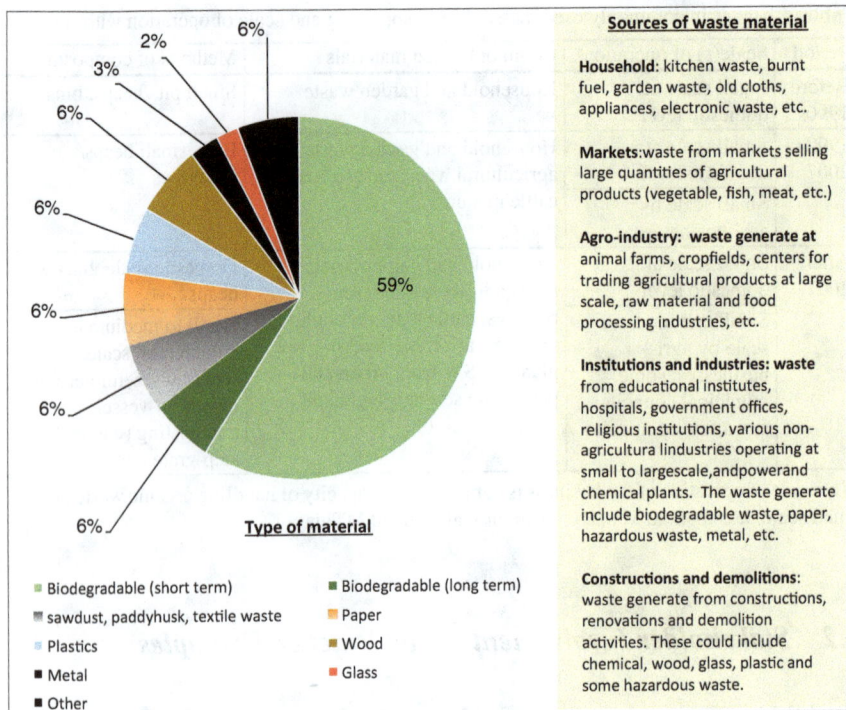

Sources of waste material

Household: kitchen waste, burnt fuel, garden waste, old cloths, appliances, electronic waste, etc.

Markets:waste from markets selling large quantities of agricultural products (vegetable, fish, meat, etc.)

Agro-industry: waste generate at animal farms, cropfields, centers for trading agricultural products at large scale, raw material and food processing industries, etc.

Institutions and industries: waste from educational institutes, hospitals, government offices, religious institutions, various non-agricultura lindustries operating at small to largescale,andpowerand chemical plants. The waste generate include biodegradable waste, paper, hazardous waste, metal, etc.

Constructions and demolitions: waste generate from constructions, renovations and demolition activities. These could include chemical, wood, glass, plastic and some hazardous waste.

Fig. Composition of solid waste generated in Sri Lanka. (Developed based on information from Arachchi 2016 and Samarasinha et al. 2015)

- Using skilled labour in the composting centre – workers are required to have a certificate from the National Vocational Training Institute. This improved the social recognition and dignity of the workers
- Addition of biochar to improve the quality of compost
- Regular quality testing of the produce
- Managing the quality and quantity of the leachate released to the environment – the centre uses diluted leachate to wet windrows and by that reduced the quantity released to the environment. Further leachate is treated before release
- Taking actions to ensure safety of workers – the centre provides safety gears to wear during compost production and performs regular health checks of the workers

So far, the compost produced has been marketed successfully although some improvement is needed in this regard.

Weligama Composting Centre

This facility was established with funds from the Ministry of Provincial Government and a few NGOs. It is being operated successfully using similar strategies as in the Balangoda composting centre (Samarasinha et al. 2015; Sinnathamby et al. 2016).

Both composting centres in Balangoda and Weligama were well-accepted by the people in those regions. These centres generated income that helped to improve infrastructure facilities in the respective areas and created a positive impact on the social and economic wellbeing of the people. Similar characteristics are seen in other successfully operating composting centres associated with local authorities (Samarasinha et al. 2015; Sinnathamby et al. 2016).

These facilities can be further improved in relation to the quality and safety of compost produced and establishment of marketing linkages with end users. In addition to biodegradable agro-industry waste, night soil that comprise of human faeces collected from septic tanks, slaughterhouse wastes, and animal farm wastes are being collected and used for compost production in these centres. Septic waste, farm waste, and slaughterhouse waste are though potential sources of contaminants such as pathogens. The quality and safety of compost produced can be improved if the pathogen-free waste stream is separated from the highly pathogen-contaminated waste stream when producing compost (Hamer 2003). This would enable the establishment of separate treatment steps to handle pathogen-contaminated waste prior to its use in compost preparation or being released directly to the environment facilitating safe disposal. More policy-level interventions are needed, and facilities should be developed with public-private partnership to regulate safe disposal and/or recycling of electronic waste which is the source of most of the toxic trace elements in solid waste (Sinnathamby et al. 2016; Franchetti 2017).

4　Safety Concerns Raised by the Major Raw Materials Used

Due to the push for recycling solid waste and a poor understanding of quality and safety concerns, many commercial-scale compost producers tend to use any biodegradable material that they can find conveniently in their composters. Constituents of compost are largely determined by what is being used as the raw material during production. Any biodegradable material could be composted. But if source materials and the composting process are not carefully regulated, whether the end product is going to meet the quality and safety standards cannot be guaranteed. Table presents some of the most popular materials used in compost making. All the raw materials listed therein are being used to different degrees by the composters in Sri Lanka. Among these, poultry manure and MSW are the most widely used raw material by large-scale composters in Sri Lanka. The heterogeneity of unsorted waste used for composting at medium- and large-scale operations poses a great challenge for managing safety and hygiene of the process as well as the quality and safety of the compost produced (Samarasinha et al. 2015). Poultry manure is also a popular material used in composting by small- to medium-scale compost producers. The

Table Widely used raw materials for compost making and concerns raised due to composition

Raw material(s)	Major concern(s)	References
Municipal solid waste	MSW contains mixture of waste materials containing plastics and hazardous waste in addition to biodegradable organic waste. Compost could have potentially toxic trace elements, micro- and nano-plastics, antibiotic resistance determinants, pathogens, bio-aerosols, and inorganic and organic pollutants. These constituents could harm workers during compost preparation and easily spread in the environment during composting	Smith (2018), Riber et al. (2014), and Samarasinha et al. (2015)
Animal manure	Farmyard wastes could include used bedding materials, faeces, and feed residues. These could harbour antibiotic resistance determinants (chemical residues and genetic information required to acquire resistance), pathogens, and parasites of animals and humans and could contaminate compost depending on characteristics of composting process. These constituents could harm workers during compost preparation and easily spread in the environment during composting	Diarrassouba et al. (2007), Dolliver et al. (2008), Furtula et al. (2010), Riber et al. (2014), Herath et al. (2015), and Herath (2017)
Crop residues and agricultural waste from markets	Crop residues collected from fields and agricultural waste could contain plant pathogens and weed seeds. Poor-quality compost prepared with these materials could lead to pest infestation	Dorahy et al. (2009)
Slaughterhouse waste	Slaughterhouse waste may contain potential human and animal pathogens. These constituents could harm workers during compost preparation and easily spread in the environment during composting	Franke-Whittle and Insam (2013)
Fly ash from power plants	Fly ash sometimes could enhance the mobility of potentially toxic trace elements in compost	Ram and Masto (2014)
Human faeces	Human faeces could harbour pathogens and parasites and antibiotic resistance determinants and organic pollutants. These constituents could harm workers during compost preparation and easily spread in the environment during composting	Pruden et al. (2013)

nature of the raw material makes this compost a cheap solution that can be made abundantly available. The major risks posed by some of the key materials used in composting are briefly discussed in the next few subsections.

4.1 Making Compost out of MSW: Opportunity Versus Challenges

Municipal solid waste is a popular choice of input in large-scale compost production in many countries, and composting MSW is often done by government-owned/ government-sponsored composting facilities and sometimes by private companies.

In the United States, solid waste collection is sub-contracted to private businesses, which saves money for the government for handling waste while promoting recycling and reusing of materials (Walls 2005). Solid waste management in many countries is practised successfully with partnerships between government, industry, and universities (Walls 2005; da Cruz et al. 2014; Franchetti 2017). In Sri Lanka, collection of MSW and composting of these materials are mainly conducted by local government authorities. The opportunities versus challenges in composting MSW in Sri Lanka are discussed in the following paragraphs considering this context.

Solid waste generation, which is estimated to be at 6,400 MT/day, has led to a number of issues with respect to resource management (Basnayake and Visvanathan 2014; Arachchi 2016). Unsafe disposal of waste, pollution of environment due to bad management practices at waste disposal sites, and unceremonious dumping of waste in improper locations have created a number of socioeconomic and health issues. In Sri Lanka, the common non-biodegradable waste materials such as glass, plastics, metals, coconut shells, and textile have a secondary market. Hence, such materials are collected separately by local vendors. Further, some municipalities collect biodegradable and non-biodegradable waste separately. They use biodegradable waste in composting, sort the non-biodegradable waste for recycling, and use the rest of it in landfills. The biodegradable waste in MSW has been observed to range between 30 and 75% in different regions; and this percentage is highly influenced by the type of waste collection systems in operation as well as by the effectiveness of waste recycling and reuse programmes (Basnayake and Visvanathan 2014; Arachchi 2016; Samarasinha et al. 2015). Waste collection by local authorities is estimated as 2,680 MT/day and constitutes about 62% of biodegradable materials on average (Basnayake and Visvanathan 2014; Arachchi 2016); thus, there is a great opportunity to recycle nutrients back to agricultural systems (Samarasinha et al. 2015). However, there are many challenges to overcome. Most of the challenges are due to the less organised MSW management structure within the country, starting with the lack of sorting. Key challenges are discussed in the following paragraphs.

According to the Municipal Council Ordinance Sections 129, 130, and 131, the Urban Council Ordinance Sections 118, 119, and 120; and *Pradesheya Saba* Act No. 15 of 1987, Sections 93 and 94, in Sri Lanka, all waste collected by local authorities including street refuse and house refuse become properties of the council, and full power to sell or dispose these materials lies with the council (Arachchi 2016). Thus, the management of solid waste has become a strict business of the local government. Local government authorities often do not have enough financial strength to develop fully equipped composting centres as income-generating entities (Sinnathamby et al. 2016). Policies and regulations are not in place to encourage private companies to engage in solid waste management.

At present, the separation of biodegradable waste according to the type at solid waste collection centres is not practised. Therefore, hazardous and biomedical waste, household waste, farm wastes, and sometimes sewage/septic waste and night soil are collected altogether as MSW (Premachandra 2006; Arachchi 2016).

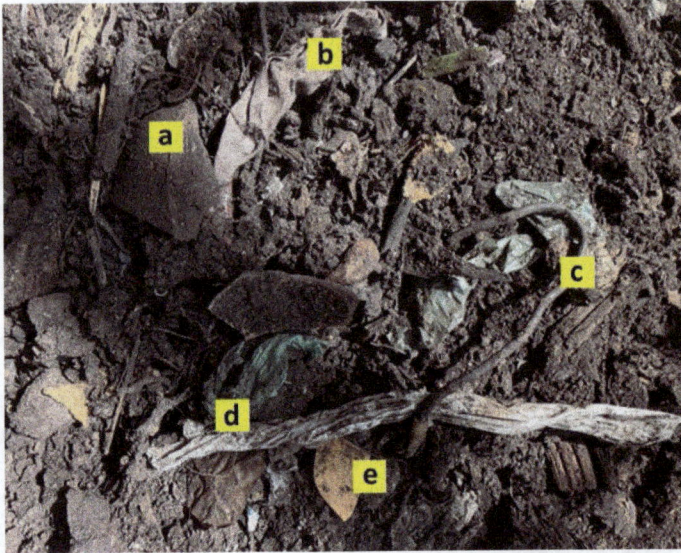

Fig. A sample of MSW-based compost just before sieving. This compost contained non-biodegradable materials such as (**a**) pericarps of fruits that are hard to decompose, (**b**) textile materials, (**c**) wires, (**d**) polythene, (**e**) leather, etc. that would end up in a rejected pile after sieving. (Photo credit: Dandeniya)

Receiving unsorted mixed waste as the starting material is the biggest challenge for composting MSW (Basnayake and Visvanathan 2014). In addition to hazardous and biomedical waste such as batteries, small electronic appliances, and hospital waste that could release potentially harmful organic and inorganic contaminants to the compost, non-biodegradable materials like plastics, textile, and polythene are also present in large volumes in MSW (Hamer 2003; Arachchi 2016). These non-biodegradable materials will increase the pile of rejected materials at the end of the composting process . Storing the rejected pile has become a great challenge for effective space management in composting facilities (Samarasinha et al. 2015).

Most of the municipalities collect waste once or twice a week mainly due to resource limitations for handling waste. Therefore, by the time waste is being collected, anaerobic digestion has started in polythene bags, which are commonly used for storing biodegradable waste (Samarasinha et al. 2015). Thus, not collecting waste during the early stage of decomposition leads to difficulties in sorting materials and handling of solid waste during composting, sometimes encouraging growth of harmful microorganisms and development of bio-aerosols and volatile chemicals that could be harmful for the handlers. Moreover, very often the waste collected by local authorities contains more moisture than the optimum level for composting. This results in production of a leachate during the composting process, which creates many problems such as pollution of surface and groundwater, skin irritation of people working in composting facilities, bad odour, and fly infestations.

Inconsistency in the quality of compost produced, safety concerns in handling materials and of the end product, high labour demand to sort materials prior to composting, and requirement of special infrastructure facilities to handle waste are some examples of challenges faced by compost producers when using moist mixed waste.

In 2017, people in the Badulla municipality in Sri Lanka protested against dumping septic tank waste to solid waste collection areas located within the city periphery in close proximity of schools and the general hospital. Collecting night soil as solid waste is a common practice in all local authorities of municipal councils in Sri Lanka and in other countries (Hamer 2003; Samarasinha et al. 2015). Not considering the suitability of land for collecting solid waste and establishing compost production units has also created several problems. The nature of land and its proximity to dense habitats are important when developing mechanisms to manage leachate, flies, and stray animals. Neglecting the views of local authorities when developing national-level strategies for waste management often leads to unsuccessful projects.

4.2 Poultry Litter/Manure as Raw Material for Composting and Its Associated Challenges

Poultry represents about one-fourth of the meat produced globally (Apata 2012). Chicken meat is a widely used protein source for human nutrition. With the growth of middle-class Sri Lankan society, a clear shift in food habits took place – from a low-protein to a high-protein content diet. Increased propaganda and accessibility of meat-based products and eggs and enhanced purchasing power of consumers blessed the poultry industry in the country with a high year-round demand. As a result, a number of poultry farms in small, medium, and large scales started in the country. In intensive production units, broiler chicken is produced in less than 6 weeks (Gerber et al. 2007); hence, one flock of birds leave the farm every 36 - 42 days, and in parallel, the bedding material containing bird droppings and feed residues should be removed. One major problem faced by the poultry farmers in the country is waste management.

In broiler farms, one popular solution is to sell or provide free dispatch farmyard waste (hereafter referred to as poultry litter) to crop farmers and compost producers (Herath et al. 2015). Practice requires that poultry litter is piled on land and allowed to mature for a couple of months and sold or used for cultivation of crops as an organic fertiliser or used as a raw material in compost production (Herath et al. 2015). This matured poultry litter is called poultry manure, or farmers sometimes refer to it as composted poultry manure although the process undertaken does not follow a proper composting technique. Poultry litter/manure is a popular low-cost raw material highly available for composting. In addition to nutrients, poultry litter/manure adds volume to the compost because the material contains paddy husk or sawdust which takes a long time to degrade and thus makes the compost available to the market on regular shorter cycles. Non-standardised manure and improperly generated compost procedures are causes of public health concerns which have not been considered in the past and that are going behind the pathogen or heavy metal concentration. Poultry litter/manure is a source of antibiotic resistance determinants and, therefore, imposes a silent threat to environmental quality and health (Herath et al. 2016). The environmental concerns from using these materials are not as visible in the short run as those from MSW. Hence, the general public and composters are mostly unaware about the challenges and threats associated with using poultry litter/manure in compost making.

The poultry industry practises genetic selection, improved feeding, and health management practices to increase production to meet the growing demand for meat and eggs. Antibiotics are extensively used in the poultry industry at sub-therapeutic dosages along with feed and water as a prophylactic method (Gerber et al. 2007; Apata 2012; Herath et al. 2015). The intensive use of antibiotics in animal husbandry leads to the establishment of a pool of antibiotic-resistant genes in the environment (Pruden et al. 2013). This has been observed in the Sri Lankan case as well (Herath et al. 2015; Herath 2017). The use of fluoroquinolone antibiotics in broiler chickens has caused an emergence of resistant *Campylobacter* in poultry (Randall et al. 2003). The extensive use of tetracycline in large-scale poultry production units in Sri Lanka has led to high levels of resistance development among bacteria rendering tetracycline no longer effective in disease control on those farms (Dandeniya et al. 2018). The tetracycline-resistant bacteria populations (10 ppm tetracycline) in poultry litter collected from those farms were as high as 9.88 ± 0.47 $\log_{10}$ CFU/g, and the enrofloxacin-resistant p opulations (10 p pm enrofloxacin) we re ar ound 9.13 ± 0.50 $\log_{10}$ CFU/g (when total culturable bacteria population size was 11.54 ± 0.49). The populations of tetracycline- and enrofloxacin-resistant bacteria (10 ppm concentration) in medium-scale poultry farms were 4.26 ± 0.46 and 2.97 ± 0.42 $\log_{10}$ CFU/g, respectively (when total culturable bacteria population size was 6.73 ± 0.52). Therefore, it is not surprising if the poultry practitioners of large-scale farms would have to abandon the use of enrofloxacin in the near future considering the level of resistance building up. Majority of the bacteria isolates obtained from these farms showed minimum inhibitory concentration and minimum bactericidal concentrations (MIC and MBC) for tetracycline greater than 128 ppm, and a significant proportion had values exceeding 1024 ppm. These results suggest an acquired resistance towards tetracycline through horizontal gene transfer events in bacteria, which has strong implications on the spread of antimicrobial resistance in the environment. The most commonly used antibiotics in broiler farms in Sri Lanka representing both medium- and large-scale operations are presented in Fig..

Most of the antimicrobials and some antibiotic-resistant gut flora end up in waste products generated from poultry production including poultry litter (Diarrassouba et al. 2007; Herath et al. 2015; Chen and Jiang 2014). Nearly 3 to 60% of antibiotics administered to an animal is excreted back through litter, and the concentrations in litter could range from 0.2 to 66 mg/kg (Diarrassouba et al. 2007; Furtula et al. 2010). Both antibiotic residues and antibiotic-resistant gene pool that end up in poultry litter could serve as antibiotic resistance determinants in the environment.

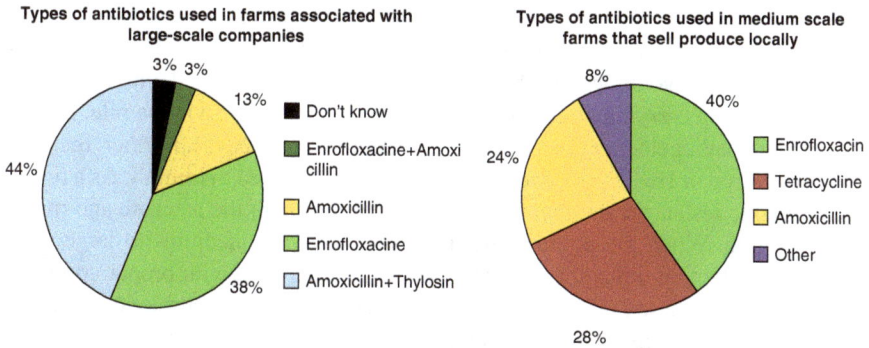

Fig. Most commonly used antibiotics in broiler farms managed in large and medium scales in Sri Lanka. (Source: Dandeniya et al. 2018)

During the composting process, microbial community structure shifts mostly due to changes in abiotic factors such as pH, temperature, and the duration it prevails at a certain range and moisture level (Kim et al. 2009). Due to heat generation in the composting process, most of the pathogens are eliminated. In addition, microbial antagonism, production of organic acids, pH change, desiccation, starvation stresses, exposure to ammonia emission, and competition for nutrients also inactivate pathogens (Wilkinson et al. 2011). Applying heat treatment after or before composting is a technique that reduces or eliminates potential bacterial pathogens in animal wastes (Chen and Jiang 2014; Herath et al. 2015).

Composting is divided into four main phases based on temperature and the active microbial community, namely, mesophilic, thermophilic, cooling, and maturation phases. In the aerobic microbial decomposition phase, heat is generated, increasing the temperature of compost mixtures to the thermophilic zone (45 to 75 °C). These temperatures can cause death of mesophilic pathogens such as *Escherichia coli* O157:H7 and *Salmonella* sp. (Ryckeboer et al. 2003). Chen and Jiang (2014) observed that during the curing of poultry litter to form poultry manure, the temperatures inside the poultry litter pile change to some extent during composting. Therefore, depending on the maturity of poultry manure, the species composition of microbial communities and, thus, the abundance of antibiotic resistance bacteria can be expected to change (Herath et al. 2015).

The survival of bacteria resistant to 100 ppm of tetracycline was observed even after subjecting broiler litter to 80 °C heat for 24 h (Herath 2017). Thus, there is a possibility that antibiotic-resistant bacteria could survive the composting process, especially when poultry litter is composted. The risk is high when the temperature rise during composting is uneven and does not reach values above 60 °C (Herath 2017). In addition to differences in relation to the maturity of manure, differences were observed in antibiotic-resistant communities with respect to intensity of management (large- to medium- to small-scale production units) and type of target production (layer farms vs. broiler farms) as well (Herath 2017; Dandeniya et al. 2018).

When poultry manure or so-called composted poultry manure is produced by farmers at small scale by just piling up the poultry litter to mature, it increases the risk of spreading antimicrobial resistance due to two reasons. Firstly, the temperature inside the pile will not reach thermophilic stage. Maturing the pile at a temperature below 40 °C would lead to the proliferation of pathogenic and enteric bacteria (high-risk group carrying antibiotic resistance genetic elements) in the pile, instead of reducing pathogenic and harmful bacteria (Herath et al. 2015). Further, the piles are not covered or sheltered; hence, they are directly exposed to rainfall. As a result, the antibiotic resistance determinants could spread through the leachate and runoff from the pile. When the poultry litter/manure is used by medium- to large-scale compost production units, the windrow system is often used with proper composting techniques. These systems could contribute to a spread in antibiotic resistance determinants in the environment if temperature regulation and leachate and runoff control have not been attended to properly.

4.3   Challenges Associated with Composting Cattle Manure

Cattle manure is traditionally used in compost preparation, especially under small-holder settings. Cattle management in Sri Lanka is mainly practised for milk production in a less intensive scale. Rearing of cattle is mostly practised free-range. Sheltered rearing is practised only in some regions, and very often those farms are operated at a small scale and scattered in the rural areas. Therefore, collecting cattle manure in adequate quantities for compost making at medium or large scales is a challenge. Only a few farms exist where cattle are reared in large scale intensively. Those farms mainly recycle cattle manure in their pasture fields. Because of these reasons, cattle manure is used as a raw material only when composting is practised in small scale.

Antibiotic use in cattle farms is less intensive as compared to that in the poultry industry. So far, no structured study has been reported in Sri Lanka on antibiotic-resistant bacteria introduced by cattle manure to compost nor has the potential for surviving the abiotic composting process been demonstrated. Nevertheless, this does not mean that the risk might not exist for cattle manure. In other countries, *E. coli* was instead found as the bacteria that better resisted the thermic temperatures, especially if the time for compost maturation was not respected. Tetracycline resistance genes were the genes most commonly found, and studies (Sharma et al. 2009) revealed that the gene levels are not affected by the composting process. It would be indeed important for risk assessment to identify the resistant bacteria surviving the composting process and their survival in the environment following land application.

4.4 Human Faeces as a Source of Raw Material in Composting

Night soil and septic waste are often dumped at MSW collecting locations in some municipalities along with other waste, without any separation (Samarasinha et al. 2015). Thus, human faeces have automatically become a raw material in the MSW composting process. Historical records indicate the use of human faeces in crop nutrient management in ancient Egyptian and Chinese civilisations. However, no such records exist in traditional agriculture in Sri Lanka. Social reception towards products that use human faeces as a raw material remains poor in the country. At present, the possibility of producing compost using human faeces is being researched by a group of scientists in Wayamba University of Sri Lanka with the financial support from the International Water Management Institute (IWMI 2018). The researchers seek to address the problem of managing sewage waste generated from households and unsafe disposal of night soil and septic waste. However, organic pollutants introduced with detergents and antibiotic resistance determinants and pathogens surviving in night soil and septic waste and the fate of these constituents during composting have not been studied extensively in Sri Lanka and would require more attention in the future.

Considering the popularity and availability of antibiotics for managing diseases, it is important to investigate the development of antibiotic resistance through all possible means. Recent medical surveys in the country indicated that one in three antibiotic prescriptions given to patients is an unnecessary directive, contributing to widespread misuse of antibiotics. Antibiotics such as amoxicillin, ampicillin, and Augmentin can be easily purchased from drugstores. Although there are regulations requiring prescriptions, these are not being practised. The misuse of common antibiotics is very high. An increase in antibiotic resistance development and the occurrence of multiple drug resistance have been observed in hospital settings (Liyanapathirana and Thevanesam 2016). This is an area that needs due attention since almost all municipalities producing compost using solid waste in Sri Lanka are not segregating materials based on their biological safety. While improving the composting process to lower the risks of spreading pathogens and antibiotic resistance determinants with the use of human faeces, attention should also be given to the study of the safety of the production process and the final product itself. Therefore, quality assessments should go beyond the set standards for compost. A revision of quality standards for compost based on the risk of resistant pathogens and antimicrobial resistance spread in the environment or risks for food chain contamination is recommended when scientific evidence reaches a critical mass. For the moment, precautionary measure should be undertaken, especially in the frame of food chain production.

5 Environmental Impact: Status and the Way Forward

The quality or the environmental significance of compost is a relative concept that should be viewed considering the target application. The primary objective of applying compost to crop fields is to supply nutrients; thus, it is considered as an organic fertiliser. In addition, there can be different objectives such as use as landfill material, nutrient source for plants in reclamation sites/non-food crops, and use as soil amendment to improve long-term soil fertility aspects (Hurst et al. 2005; Smith 2009; Agegnehu et al. 2015). When producing compost to achieve these secondary objectives, the nature of materials used and quality and safety concerns during production and later use would vary. Hence, to better evaluate quality concerns, a tailor-made set of regulations and quality standards should be made available separately for the composting process and the produce targeted for different purposes (e.g. organic fertiliser, soil fertility enhancer, landfill material, fertiliser for reclamation site, etc.). This can be achieved via identifying a set of common core standards and regulations across all the targeted applications and specific conditions to relax or fulfil when the target end use is factored in. This is already being practised in some countries. A good-quality compost should be able to support plant growth by supplying nutrients and not pollute the environment by any means. It should be adequately matured, should contain nutrients and moisture in acceptable levels, should be free from pathogens and weed seeds, and should have acceptable shelf life. The acceptable values could vary from country to country and also be based on the targeted application. Presently in Sri Lanka, there is only a single set of stan-dards for compost irrespective of the intended end use .

5.1 Impact on Soil Environment

Compost contains chemicals (nutrients as well as pollutants) in different forms ranging from simple inorganic ions to complex organic forms contributing to different pools of nutrients in soil, namely, plant-available pool in soil solution, readily exchangeable pool, and reserves. Compost releases nutrients to the plant-available pool slowly at a pace close to the rate of plant nutrient uptake. Therefore, unlike chemical fertilisers, that enrich the plant-available pool for a short term, compost supports retention of nutrients in the soil system avoiding leakage. Depending on the type of compost, its contribution to different pools of nutrients in soils varies (Agegnehu et al. 2015). While compost produced with the objective of using it as a soil amendment to improve long-term soil fertility contributes mostly to nutrient reserves, compost produced as an organic fertiliser contributes mostly to the exchangeable fraction . The same phenomena hold true for the pollut-ants in compost, and some pollutants introduced to soil via compost could move through the food chain, while some other forms could remain in soil for a long duration. In addition to chemical constituents, compost also contains organisms

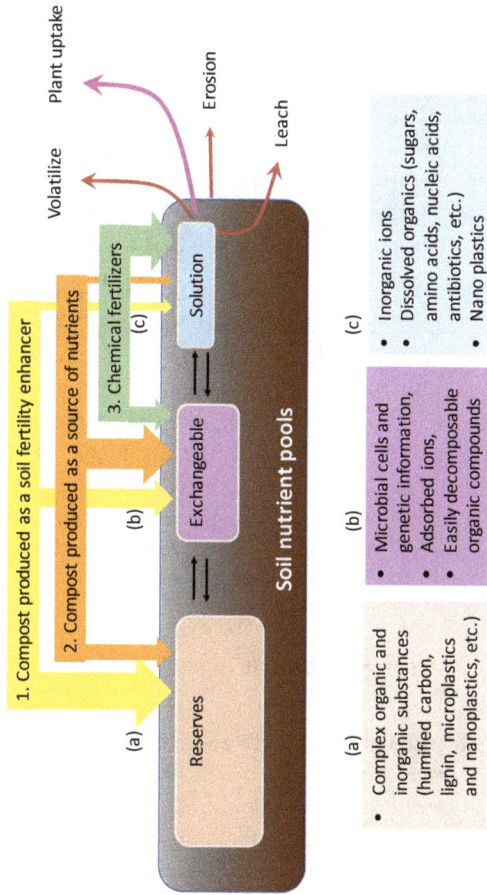

Fig. Contribution of two different types of compost (soil fertility enhancer type and organic fertiliser type) and chemical fertilisers to different nutrient pools (reserves, exchangeable, and soil solution) in soil. The width of the arrows (**a**), (**b**), and (**c**) indicate the approximate additions of constituents to each pool (bacteria, fungi, archaea, protozoans, ciliates, earthworms, etc.). It should be noted that while compost promotes nutrient cycling and benefits crop growth, some organic and inorganic constituents introduced via compost could build up in soil imposing threats to the food chain through contamination and to the overall quality of the environment (Smith 2009, 2018). The application of compost to soil over a long period creates long-lasting positive and/or negative impressions on soil characteristics. Therefore, the safety concerns related to compost should be viewed holistically from handling of raw materials to production of compost to the consequences of end applications of the product.

Some contaminants in raw materials such as microbial pathogens and parasites could spread in the environment via animals such as flies and dogs in composting sites. Bio-aerosols and volatile compounds could enable transmission from composting sites to other environments with wind. In addition, leachates coming out from compost piles during the production process and runoff water from composting sites could contaminate surface water and groundwater and may cause serious health effects on people working on-site and off-site (Samarasinha et al. 2015).

In Sri Lanka, regulations for ensuring quality of compost with respect to inorganic constituents are in place and also being implemented. The chemical and phys-ical parameters as mentioned in Table are being monitored in the compost sold on the market. However, there are restrictions related to the capacity of regulatory bodies to satisfactorily survey the market to ensure product quality. However, policymakers and the general public are aware of the importance of the identified parameters in Table for safeguarding human health and environmental quality. The following sections provide a brief overview of the emerging pollutants in com-post, of which there is very limited awareness in Sri Lanka.

5.2   Pollutants of Emerging Concern

Most of the existing quality standards already address the maximum permissible levels of potentially toxic trace elements. However, very little attention is given to the organic pollutants in compost. The wide variety of substances, continuous changes in their structures and chemical natures due to biodegradation, lack of understanding on intermediary metabolites, and technological issues in identifying these compounds and their intermediary metabolites make organic pollutants a complex field to address when developing quality standards. Organic pollutants are commonly found in MSW-based compost. Bio-aerosols and volatile compounds are generally organic. In addition, micro- and nanoplastics as well as antimicrobial resistance determinants can be considered as emerging pollutants that could be introduced to the environment through compost and get transmitted into the food chain imposing threats on human and animal health.

Organic Pollutants and Bio-aerosols

There is very limited information available on organic pollutants and bio-aerosols in compost despite the fact that scientific bodies have expressed their concern over the lack of monitoring of these compounds in compost as well as in composting facilities (Samarasinha et al. 2015). The leachate coming from MSW composting and dumping sites in Sri Lanka has been studied by a few groups (Gunawardana et al. 2009; Basnayake and Visvanathan 2014; Samarasinha et al. 2015). Gunawardana et al. (2009) observed that composting of MSW has reduced the organic pollutant content in leachate by seven times when compared to the leachate coming from untreated MSW. However, they have not gone into detail as to identify the organic pollutants in the leachate or in the compost itself. Research activities are ongoing to study the composition of gaseous emissions during composting and identifying potential organic pollutants in compost and their mobility in the environment. The importance of studying the impact of bio-aerosols (both chemical- and organism-based) released to the environment during the composting process on the health of workers at composting facilities in Sri Lanka is highlighted by Samarasinha et al. (2015). There is mounting evidence in literature on safety concerns and health risks associated with organic pollutants and bio-aerosols in compost and during its production and effects on end consumers and workers in other parts of the world (Fischer et al. 1999; Pearson et al. 2015).

Microplastics

Most of the microplastic- and nanoplastic-related studies are restricted to aquatic ecosystems, and only a very few concentrate on terrestrial environments. Although the discovery of microplastics and nanoplastics is new, these substances have entered Earth's ecosystems as early as the 1900s with commercial production and use of plastics around the world. Microplastics and nanoplastics in compost originate from MSW containing exfoliates in cosmetics and/or partially degraded larger plastic debris, including polyester fibres from fabrics, polyethylene fragments from plastic bags, plastic cans, and parts of electronic appliances (Galloway 2015). Plastic waste is common in MSW in Sri Lanka and contributes to a significant volume in the rejected pile after composting .

Smith (2018) detected microplastics in MSW-based compost. Based on the observations so far, microplastics and nanoplastics, once added to soil, could influence micro-fauna and affect the soil food web (Rillig 2012). Changes in the dynamics of microbial communities affect soil properties leading to a chain of reactions influencing the ecosystem functions. Smith (2018) observed that microplastics introduced via compost could accumulate in plant tissues. Since micro- and nano-sized plastics can bioaccumulate, there is a risk for human and animal health. As this is an emerging area of study, direct clinical evidences are lacking. However, potential influences of microplastics and nanoplastics on immune response-related disorders and cancer incidents are being suggested (Galloway 2015; Wright and

Kelly 2017). Due to difficulties in studying the material, microplastic- and nanoplastic-related research is not yet popular. However, a few research programmes are being conducted on this group of pollutants as summarised in Galloway (2015) and Wright and Kelly (2017). There is no reported literature on the presence of microplastics and nanoplastics in compost and their possible consequences in Sri Lanka. Therefore, this is an area of research that must be initiated urgently, especially considering the wide use of compost and rejected piles from MSW-handling authorities for land management.

Antibiotic Resistance Determinants

Antibiotics are the most successful and sometimes the only treatment against infectious diseases caused by bacteria and fungi. However, the spread of antibiotic resistance traits among pathogenic groups of microorganisms is challenging the healthcare systems worldwide including Sri Lanka (Martinez 2009; Pruden et al. 2013; Liyanapathirana and Thevanesam 2016). Antibiotic residues, antibiotic-resistant bacteria, and genetic information containing antibiotic resistance genes (free DNA) are considered as antibiotic resistance determinants (Martinez 2009). While antibiotic residues exert a selection pressure, the organisms having the trait and the free DNA containing resistance genes serve as the genetic pool that facilitate horizontal gene transfer (Riber et al. 2014), speeding up the evolution of antibiotic resistance traits in a microbial community. The antibiotic resistance determinants are abundant in human and animal faecal material (Pruden et al. 2013), which is sometimes used as raw material for compost making as discussed in the previous section. Animal manure could harbour pathogenic microorganisms such as *Salmonella* spp., *Listeria monocytogenes*, and *Escherichia coli* O157:H7 that could contaminate fresh produce from crop lands and cause foodborne illnesses (Johannessen et al. 2004; Maffei et al. 2013). These organisms with enteric origin could easily develop multiple drug resistance traits. Although composting is being viewed as a method to reduce antibiotics in animal manure, some recent literature indicates that the degradation of antibiotics during composting is not uniform across different classes of antibiotics (Dolliver et al. 2008). Considering the large proportion of animal manure used by the composting industry in Sri Lanka, it is important to be cautious about the spread of antibiotic resistance traits along the food chain as a result of handling and processing of animal manures and human faeces.

5.3   The Way Forward

At present, the Central Environmental Authority of Sri Lanka is the government body involved in conducting environmental impact assessments for all industries operating at commercial scale. They should play an active role in ensuring environmental quality and safety in relation to composting facilities. The Sri Lanka Institute of Standards and the Consumer Affairs Authority are involved with ensuring the quality and safety of products in the market. Therefore, they have a major role to play to ensure the quality of compost. The knowledge gap with respect to pollutants of emerging concern and quality control of compost based on targeted end applications needs to be addressed by these authorities (as the authorities having the mandate to safeguard consumers and the environment) in collaboration with the scientific community in the country.

Efforts taken in the past to popularise compost production and use have not reached the expected outcomes mainly due to the lack of involvement of key stakeholder groups in project development and the decision-making process. The relevant government and corporate sector groups need to be consulted when developing strategies to improve the quality of compost as they are very aware of the unique challenges at ground level. It is also important to emphasise the need for obtaining the support of the scientific community, who can address the gaps, to ensure safe composting in the country. There should be proper coordination between the stakeholder groups to safeguard the environment against the spread of pollutants due to composting and the use of compost in agriculture and landfills. Funding agencies such as the National Science Foundation (NSF), National Research Council (NRC), and Sri Lanka Council for Agricultural Research Policy (SLCARP) could foster cooperation through financial support for projects done in a multidisciplinary nature addressing the key challenges previously presented in this chapter related to safe composting in Sri Lanka.

6 Conclusions

Compost production in Sri Lanka is happening at different scales and with a wide variety of biodegradable waste as raw material. The risks and limitations associated with the composting process and the quality of the end product are highly influenced by the scale of operation and the nature of raw materials used. MSW is a raw material with good potential for use in compost production, but there are a number of challenges to be addressed to overcome the negative impact on the environment and public health. One major aspect affecting the quality of MSW compost is the inadequate sorting/segregation of MSW used for compost production. MSW and poultry manure/litter, which are some common raw materials used in medium- to large-scale composting facilities in Sri Lanka, contribute to the spread of some pollutants of emerging concerns. Antimicrobial resistance determinants, micro- and nanoplastics, and organic pollutants are major emerging pollutants that have not yet received the attention of policymakers and regulatory bodies in Sri Lanka.

References

Agegnehu, G., Bass, A. M., Nelson, P. N., Muirhead, B., Wright, G., & Bird, M. I. (2015). Biochar and biochar-compost as soil amendments: Effects on peanut yield, soil properties and greenhouse gas emissions in tropical North Queensland, Australia. *Agriculture, Ecosystems & Environment, 213*, 72–85.

Apata, D. F. (2012). The emergence of antibiotics resistance and utilization of probiotics for poultry production. *Science Journal of Microbiology, 2*, 8–12.

Arachchi, K. H. M. (2016). *Present status of solid waste management and challenges for change.* Colombo: National solid waste management program in Sri Lanka. Central Environmental Authority.

Basnayake, B. F. A., & Visvanathan, C. (2014). Solid waste management in Sri Lanka. In *Municipal solid waste management in Asia and the Pacific Islands* (pp. 299–316). Singapore: Springer.

Chen, Z., & Jiang, X. (2014). Microbiological safety of chicken litter or chicken litter-based organic fertilizers: A review. *Agriculture, 4*(1), 1–29.

da Cruz, N. F., Ferreira, S., Cabral, M., Simões, P., & Marques, R. C. (2014). Packaging waste recycling in Europe: Is the industry paying for it? *Waste Management, 34*(2), 298–308.

Dandeniya, W. S., Herath, E. M., Kasinthar, M., Lowe, W. A. M., & Jinadasa, R. N. (2018). Prevalence of antibiotic resistant bacteria in poultry litter based manures and potential threats on food safety for carrot. In *Global symposium on soil pollution*. Rome: Food and Agriculture Organization.

Déportes, I., Benoit-Guyod, J. L., & Zmirou, D. (1995). Hazard to man and the environment posed by the use of urban waste compost: A review. *Science of the Total Environment, 172*(2–3), 197–222.

Diarrassouba, F., Diarra, M. S., Bach, S., Delaquis, P., Pritchard, J., Topp, E., & Skura, B. J. (2007). Antibiotic resistance and virulence genes in commensal Escherichia coli and Salmonella isolates from commercial broiler chicken farms. *Journal of Food Protection, 70*, 1316–1327.

Dolliver, H., Gupta, S., & Noll, S. (2008). Antibiotic degradation during manure composting. *Journal of Environmental Quality, 37*(3), 1245–1253.

Dorahy, C. G., Pirie, A. D., McMaster, I., Muirhead, L., Pengelly, P., Chan, K. Y., Jackson, M., & Barchia, I. M. (2009). Environmental risk assessment of compost prepared from Salvinia, Egeria densa, and Alligator weed. *Journal of Environmental Quality, 38*(4), 1483–1492.

Fischer, G., Schwalbe, R., Möller, M., Ostrowski, R., & Dott, W. (1999). Species-specific production of Microbial Volatile Organic Compounds (MVOC) by airborne fungi from a compost facility. *Chemosphere, 39*(5), 795–810.

Franchetti, M. J. (2017). University waste reduction and pollution prevention assistance programs: Collaborations with industry, government and academia. In *Sustainability practice and education on university campuses and beyond* (pp. 50–59). Sharjah: Bentham Science Publishers.

Franke-Whittle, I. H., & Insam, H. (2013). Treatment alternatives of slaughterhouse wastes, and their effect on the inactivation of different pathogens: A review. *Critical Reviews in Microbiology, 39*(2), 139–151.

Furtula, V., Farrell, E. G., Diarrassouba, F., Rempel, H., & Diarra, M. S. (2010). Veterinary pharmaceuticals and antibiotic resistance of Escherichia coli isolates in poultry litter from commercial farms and controlled feeding trials. *Poultry Science, 89*, 180–188.

Galloway, T. S. (2015). Micro- and nano-plastics and human health. In *Marine anthropogenic litter* (pp. 347–370). Berlin: Springer.

Garcıa-Gil, J. C., Plaza, C., Soler-Rovira, P., & Polo, A. (2000). Long-term effects of municipal solid waste compost application on soil enzyme activities and microbial biomass. *Soil Biology and Biochemistry, 32*(13), 1907–1913.

Gerber, P., Opio, C., & Steinfeld, H. (2007). Poultry production and the environment-a review. *Animal production and health division, food and agriculture organization of the United Nations*. FAO Publishing web.

Gunawardana, E. G., Basnayake, B. F., Shimada, S., & Iwata, T. (2009). Influence of biological pre-treatment of municipal solid waste on landfill behaviour in Sri Lanka. *Waste Management & Research, 27*(5), 456–462.

Hamer, G. (2003). Solid waste treatment and disposal: Effects on public health and environmental safety. *Biotechnology Advances, 22*(1–2), 71–79.

Herath, E. M. (2017). *Investigating the effect of poultry manure application on antibiotic resistance of bacteria in intensively cultivated soils.* M.Phil. Thesis. Postgraduate Institute of Agriculture, University of Peradeniya, Sri Lanka.

Herath, E. M., Dandeniya, W. S., Samarasinghe, A. G. S. I., Bandara, T. P. M. S. D., & Jinadasa, R. N. (2015). A preliminary investigation on methods of reducing antibiotic resistant bacteria in broiler litter in selected farms in mid country Sri Lanka. *Tropical Agricultural Research Journal, 26*(2), 412–417.

Herath, E. M., Palansooriya, A. G. K. N., Dandeniya, W. S., & Jinadasa, R. N. (2016). An assessment of antibiotic resistant bacteria in poultry litter and agricultural soils in Kandy District, Sri Lanka. *Tropical Agricultural Research Journal, 27*(4), 389–398.

Hurst, C., Longhurst, P., Pollard, S., Smith, R., Jefferson, B., & Gronow, J. (2005). Assessment of municipal waste compost as a daily cover material for odour control at landfill sites. *Environmental Pollution, 135*(1), 171–177.

IWMI. (2018). *Human waste reuse could benefit farmers and improve public health in South Asia.* International Water Management Institute, Sri Lanka.

Johannessen, G. S., Froseth, R. B., Solemdal, L., Jarp, J., Wasteson, Y., & Rorvik, L. M. (2004). Influence of bovine manure as fertilizer on the bacteriological quality of organic iceberg lettuce. *Journal of Applied Microbiology, 96*(4), 787–794.

Kendaragama, K. M. A. (2006). Fertilizer use efficiency in farming systems in Sri Lanka. *Journal of Soil Science Society of Sri Lanka, 8*, 1–18.

Kendaragama, K. M. A., Lathiff, M. A., & Chandrapala, A. G. (2001). Impact of vegetable cultivation on fertility status of soils in the Nuwara Eliya area. *Annals of the Sri Lanka Department of Agriculture, 3*, 95–100.

Kim, J., Luo, F., & Jiang, X. (2009). Factors impacting the regrowth of Escherichia coli O157:H7 in dairy manure compost. *Journal of Food Protection, 72*(7), 1576–1584.

Liyanapathirana, V. C., & Thevanesam, V. (2016). Combating antimicrobial resistance. *Sri Lankan Journal of Infectious Diseases, 6*(2), 72–82.

Maffei, D. F., Silveira, F. A. N., & Mortatti Catanozi, M. P. L. (2013). Microbiological quality of organic and conventional vegetables sold in Brazil. *Food Control, 29*(1), 226–230.

Mapa, R. B. (2003). Sustainable soil management in the 21st century. *Tropical Agricultural Research and Extension, 6*, 44–48.

Martinez, J. L. (2009). Environmental pollution by antibiotics and by antibiotic resistance determinants. *Environmental Pollution, 157*(11), 2893–2902.

Nagarajah, S. (1986). Fertilizer recommendations for rice in Sri Lanka: A historical review. *Journal of Soil Science Society of Sri Lanka, 4*, 4–14.

Palm, O., & Sandell, K. (1989). Sustainable agriculture and nitrogen supply in Sri Lanka: farmers' and scientists' perspective. *Ambio, 18*(8), 442–448.

Pearson, C., Littlewood, E., Douglas, P., Robertson, S., Gant, T. W., & Hansell, A. L. (2015). Exposures and health outcomes in relation to bioaerosol emissions from composting facilities: A systematic review of occupational and community studies. *Journal of Toxicology and Environmental Health Part B, 18*(1), 43–69.

Premachandra, H. S. (2006). Household waste composting & MSW recycling in Sri Lanka. In *Asia 3R conference*, Tokyo.

Pruden, A., Larsson, D. J., Amézquita, A., Collignon, P., Brandt, K. K., Graham, D. W., Lazorchak, J. M., Suzuki, S., Silley, P., Snape, J. R., & Topp, E. (2013). Management options for reducing the release of antibiotics and antibiotic resistance genes to the environment. *Environmental Health Perspectives, 121*(8), 878–885.

Ram, L. C., & Masto, R. E. (2014). Fly ash for soil amelioration: A review on the influence of ash blending with inorganic and organic amendments. *Earth-Science Reviews, 128*, 52–74.

Randall, L. P., Ridley, A. M., Cooles, S. W., Sharma, M., Sayers, A. R., Pumbwe, L., & Woodward, M. J. (2003). Prevalence of multiple antibiotic resistance in 443 Campylobacter spp. Isolated from humans and animals. *Journal of Antimicrobial Chemotherapy, 52*(3), 507–510.

Riber, L., Poulsen, P. H., Al-Soud, W. A., Skov Hansen, L. B., Bergmark, L., Brejnrod, A., Norman, A., Hansen, L. H., Magid, J., & Sørensen, S. J. (2014). Exploring the immediate and long-term impact on bacterial communities in soil amended with animal and urban organic waste fertilizers using pyrosequencing and screening for horizontal transfer of antibiotic resistance. *FEMS Microbiology Ecology, 90*(1), 206–224.

Rillig, M. C. (2012). Microplastic in terrestrial ecosystems and the soil? *Environmental Science & Technology, 46*(12), 6453–6454.

Ryckeboer, J., Mergaert, J., Coosemans, J., Deprins, K., & Swings, J. (2003). Microbiological aspects of biowaste during composting in monitored compost bin. *Journal of Applied Microbiology, 94*(1), 127–137.

Samarappuli, L. (2001). Nutrition. *Handbook of rubber, volume 1: Agronomy* (pp. 156–175). Rubber Research Institute of Sri Lanka, Sri Lanka.

Samarasinha, G. G. D. L. W., Bandara, M. A. C. S., & Karunarathna, A. K. (2015). *Municipal solid waste composting: Potentials and constraints* (HARTI research report No: 174). Colombo: Hector Kobbekaduwa Agrarian Research and Training Institute.

Sanchez, P. A. (1997). Changing tropical soil fertility paradigms: From Brazil to Africa and back. In *Plant-soil interactions at low pH* (pp. 19–28). Piracicaba: Brazilian Society of Soil Science.

Sharma, R., Larney, F. J., Chen, J., Yanke, L. J., Morrison, M., Topp, E., Tim, A., McAllister, A., & Yu, Z. (2009). Selected antimicrobial resistance during composting of manure from cattle administered sub-therapeutic antimicrobial. *Journal of Environmental Quality, 38*, 567–575.

Sinnathamby, V., Paul, J. G., Dasanayaka, S. W. S. B., Gunawardena, S. H. P., & Fernando, S. (2016). Factors affecting sustainability of municipal solid waste composting projects in Sri Lanka. *1st International Conference in Technology Management (iNCOTeM)* (p. 98).

Smith, S. R. (2009). A critical review of the bioavailability and impacts of heavy metals in municipal solid waste composts compared to sewage sludge. *Environment International, 35*(1), 142–156.

Smith, M. (2018). *Do microplastic residuals in municipal compost bioaccumulate in plant tissue?* M.Sc. Thesis. Faculty of Social and Applied Sciences, Royal Roads University, British Columbia, Canada.

Sri Lanka Standards Institution (SLSI). (2003). Specifications for municipal solid waste and agricultural waste. *Technical Guidelines on Solid Waste Management in Sri Lanka, SLSI 1246:2003* (pp. 39–51). Central Environmental Authority, Sri Lanka.

Tennakoon, N. A., & Bandara, S. H. (2003). Nutrient content of some locally available organic materials and their potential as alternative sources of nutrients for coconut. *Cocos, 15*, 23–30.

TRI advisory circular. (2000). Fertilizer recommendations for mature tea. *Circular No. SP3. Serial No. 00/3*. Tea Research Institute. Sri Lanka.

TRI advisory circular. (2016). Fertilizer recommendations for mature tea in small holdings. *Interim circular No. SP10. Serial No. 01/16*. Tea Research Institute, Sri Lanka.

Vision 2025 – A country Enriched. (2017). *The changing face of a dynamic modern economy*. Prime Minister's Office, Sri Lanka.

Walls, M. (2005). How local governments structure contracts with private firms: Economic theory and evidence on solid waste and recycling contracts. *Public Works Management and Policy, 9*, 206–222.

Weerahewa, J., Kodithuwakku, S. S., & Ariyawardana, A. (2010). The fertilizer subsidy program in Sri Lanka. In *Food policy for developing countries: Case studies*. Ithaca/New York: Cornell University.

Weerasinghe, P. (2017). Best practices of integrated plant nutrition system in Sri Lanka. *Best practices of integrated plant nutrition system in SAARC Countries* (pp. 135–160). Dhaka SAARC Agriculture Centre.

Wijetunga, C. S., & Saito, K. (2017). Evaluating the fertilizer subsidy reforms in the rice production sector in Sri Lanka: A simulation analysis. *Advances in Management and Applied Economics, 7*(1), 31–51.

Wijewardena, J. D. H. (2005). Improvement of plant nutrient management for better farmer livelihood, food security and environment in Sri Lanka. In Improving plant nutrient management for better farmer livelihoods. In *Proceedings of the workshop on food security and environmental sustainability* (pp. 73–93). Beijing: FAO Publishing Web.

Wilkinson, K. G., Tee, E., Tomkins, R. B., Hepworth, G., & Premier, R. (2011). Effect of heating and aging of poultry litter on the persistence of enteric bacteria. *Poultry Science, 90*(1), 10–18.

Wright, S. L., & Kelly, F. J. (2017). Plastic and human health: A micro issue? *Environmental Science & Technology, 51*(12), 6634–6647.

7

Co-composting: An Opportunity to Produce Compost with Designated Tailor-Made Properties

Laura Giagnoni, Tania Martellini, Roberto Scodellini, Alessandra Cincinelli, and Giancarlo Renella

Abstract Co-composting is a technique that allows the aerobic degradation of organic waste mixtures, primarily aiming at obtaining compost that can be used as fertiliser or soil amendment. As compared to the typical composting activity, the main difference is not merely the use of more than one feedstock to start and sustain the biodegradation process, but also the possibility of combining various kinds of waste to obtain 'tailored' products with designed properties, or to reclaim and valorise natural resources, such as degraded soils or polluted soils and sediments. Set up of appropriate co-composting protocols can be a way to optimise the management of waste produced by different sectors of agriculture and industry and also from human settlements. Different formulations can not only optimise the biodegradation process through the adjustment of nutrient ratios, but also lead to the formation of products with innovative properties. Moreover, co-composting can be a technique of choice for the reclamation of soils degraded by intensive agriculture or contaminated soils and sediments. In fact, an appropriate mix of organic waste and soils can restore the soil structure and induce fertility in nutrient-depleted soils, and also remediate polluted soils and sediments through degradation of organic pollutants and stabilisation of heavy metals. While the selection of different mixes of organic waste may lead to the design of composts with specific properties and the potential valorisation of selected waste materials, there are still several factors that hamper the development of co-composting platforms, mainly insufficient knowledge of

L. Giagnoni
Department of Agriculture, Food, Environment and Forestry (DAGRI), University of Florence, Florence, Italy
e-mail: laura.giagnoni@unifi.it

T. Martellini · R. Scodellini · A. Cincinelli
Department of Chemistry, University of Florence, Sesto Fiorentino, Italy
e-mail: tania.martellini@unifi.it; roberto.scodellini@unifi.it; alessandra.cincinelli@unifi.it

G. Renella (✉)
Department of Agronomy, Food, Natural Resources, Animals and Environment, University of Padua, Legnaro, Italy
e-mail: giancarlo.renella@unipd.it

some chemical and microbiological processes, but also some legislative aspects. This chapter illustrates the progress achieved in co-composting technology worldwide, some key legislative aspects related to the co-composting process, the main scientific and technical aspects that deserve research attention to further develop co-composting technology, and successful applications of co-composting for the reclamation of soils and sediments, allowing their use for cultivation or as growing media in plant nurseries. A specific case study of the production of fertile plant-growing media from sediment co-composting with green waste is also illustrated.

Keywords Co-composting · Product design · Waste recycle · Co-composting process evaluation · Dredged sediments · Green waste co-composting

1 Introduction

Co-composting is the process of the aerobic degradation of organic compounds using more than one feedstock. The initial materials can be of industrial, agricultural or urban domestic household origin, and all materials allowed to be recycled as bioresources according to the local legislation. To date, reports have been published on the co-composting of sewage sludge, animal excreta, urban solid waste and plant residues, from various pilot experimental activities and industrial scale treatment plants. Since ever, composting of organic solid waste was directed towards the sanitation and volume reduction of municipal waste, with extensive efforts directed to mechanical innovations, the reduction of emissions and odours during the composting process, and speeding up of compost maturation through the achievement of a sustained thermophilic phase.

Currently, with rapid urbanisation occurring globally, organic waste represents the majority of the municipal waste in emerging countries. In this context, composting systems can play an important role in managing waste as well as creating employment and creating products that contribute to food security, particularly in developing countries. Organic solid waste collected from households and institutions is composted either at decentralised (community-based) or centralised composting plants. Community-based decentralised composting systems can generally process about 2–50 tonnes per day of organic waste, whereas centralised composting facilities are capable of receiving 10–200 tonnes per day. There are two fundamental types of composting techniques: open or windrow composting, a slower process conducted outdoors with simple equipment, and enclosed system composting, where composting is performed in a building, tank, box, container or vessel. Proper management of the plant and marketing of the compost are key factors to ensure the sustainability of such systems.

The above-mentioned global trends call for new sustainable and safe strategies for waste treatment, for their recycling and for the minimisation of landfill. Research in compost science and technology has mainly focused on the treatment and

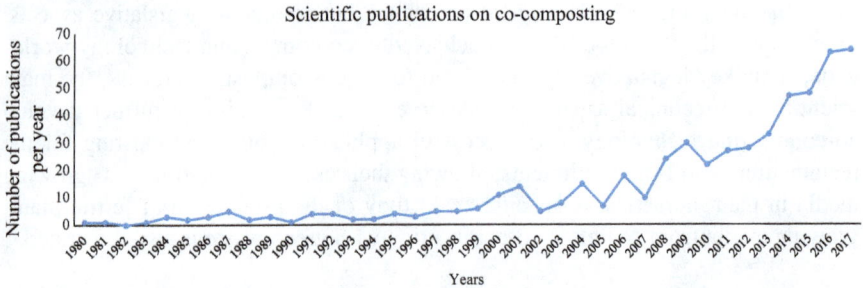

Fig Trend of publications using the term co-composting in the title or keywords in the period 1980–2018. (Source: Scopus, accessed on 30 December 2018)

conversion of the organic fraction of waste, mainly municipal solid waste and manure, which has represented the target fraction to be recycled because it is rich in nutrients that can be of particular benefit if they are reused as compost in agriculture as soil amendment or fertiliser (Bernstad and la Cour Jansen 2011). For example, pre-treated faecal sludge can be co-composted together with the solid waste, as faecal sludge has a high moisture and nitrogen content and biodegradable solid waste is high in organic C and has good bulking properties (i.e. it allows air to flow and circulate). By combining the two types of waste, the benefits of each can be used to optimise the process and the product.

Since the 1990s some new co-composting approaches have been extended to the treatment of organic waste, assisted by the use of mineral matrices, whereas co-composting techniques aiming at the reclamation of inorganic environmental matrices such as soils and sediments are still seldom reported. A literature survey in Scopus (Elsevier) revealed that 596 articles could be retrieved using 'co-composting' as search criteria in the article title, abstract and keywords, with a marked increasing trend of scientific publications from the mid-2000s onward. Very interestingly, while the early publications focusing on co-composting were mainly published by Western European and North American countries, the substantial growth of the literature body on the subject is mostly contributed by Asian countries, mainly China, India and Malaysia (Scopus, accessed on 30 September 2018), where this approach is significantly contributing to modernising the waste management in developing countries (Hoornweg et al. 2000).

2 Co-composting: Organic Waste Composted with Inorganic and Biotic Additives

The review article by Barthod et al. (2018) exhaustively illustrates how the composting of the four major categories of organic waste (food waste, green waste, municipal solid waste and sewage sludge) has been mainly conducted by mixing them with several inorganic materials and types of waste, with the aim of improving specific aspects of the composting process. Specific features of the composting process, such as sanitation, compost maturation and odour emission; pollutant concentrations and compost grade; nutrient content and availability; and greenhouse gas emissions have received different degrees of attention depending on the 'hot topics' under discussion by the scientific community and all stakeholders. Analysis of the literature conducted as mentioned above shows that the use of additives to the main composting waste has shifted from mainly sanitation and maturation aspects, which were important in the 1980s and the 1990s (in particular, sanitation effectiveness and compost maturation related to compost grade), towards a focus on nutrient content and availability and greenhouse gas emission, which have gained increasing attention from the 2000s to date. These fundamental changes can be related to the improved compost science and technology, which has led to consolidated protocols of sanitation and improved sensitivity of the analytical procedures to the use of well-sorted waste, at least in the most developed countries, which prevents the presence of excessive heavy metal concentrations in the original organic materials, and to the concomitant stringent need to optimise the nutrient recycle in agriculture and minimise the greenhouse gas emissions during the composting process to protect the atmosphere. It is important to underline that the recent literature shows that although this paradigm shift was uniform across all countries, including those where waste management is not yet optimal, research on the basic processes and on the properties of the compost has not completely stopped, especially because new additives (i.e. zeolites, biochar) are being tested. However, even in the presence of major changes in scientific approaches, the use of the term co-composting was used mainly to highlight the use of additives instead of designing new processes with different waste materials.

Composting, as a process, has proven to be effective against several organic pollutants through their complete mineralisation or conversion into less toxic substances. Atagana (2004) reported that the addition of 25% of poultry manure to a contaminated soil previously mixed with 1:1 wood chips could reduce the polycyclic aromatic hydrocarbon (PAH) concentration more than in the soil–wood chips mixture only, to levels below 1 mg/kg, mainly through the adjustment of the C:N ratio value and temperature control, which favoured microbial proliferation. Cai et al. (2007) reported that degradation of carcinogenic PAHs in secondary dewatered sewage sludge was more rapid when mixed with rice straw, particularly when inoculated with a commercial mixed microbial/biostimulant formulation. Similar results were reported by Wan et al. (2003), who compared various mixes of pig manure, sewage sludge and soybean refuse to enhance the degradation of PAHs in a polluted soil, reaching a maximum of 90% of removal with pig manure, and for removal of total petro hydrocarbons (over 99%) during a co-composting of a diesel oil-contaminated soil mixed with sewage sludge (Namkoong et al. 2001). Huang et al. (2016) reported that co-composting of dredged sediments with rice straw, vegetable residues and bran, and bioaugmented with *Phanerochaete chrysosporium*, allowed the degradation of 4-nonylphenol, a dangerous endocrine disruptor.

Aparna et al. (2008) presented a co-composting treatment for the treatment of sediments from Isnapur, Khazipally and Gandigudem lakes (India) contaminated by benzene, phenols, PAHs, and polychlorinated biphenyls (PCBs), mixing them with poultry manure, cow dung, urea, diammonium phosphate and sawdust in the following proportions: 70% polluted sediment + 5% poultry manure + 5–8% cow dung +8% sawdust + 5–6% urea + 4–8% diammonium phosphate DAP, so as to adjust the initial C:N ratio value to 30 for all treatments. The results showed that all classes of organic contaminants were significantly reduced after co-composting and all maturity indices were met after a total time of 23 weeks. Similar results were reported by Rekha et al. (2005) on contaminated sediments from lakes in Hyderabad (India) co-composted with manure and sawdust at a ratio of 2:1:2. Concerning more recalcitrant compounds, Büyüksönmez et al. (1999) reported that organo-chlorinated compounds are highly resistant to biodegradation—even more than common pesticides. However, the efficiency of the degradation of chlorophenols in composting polluted soils was reported to be higher than 80% (Bentham and McClure 2003) and 90% (Laine and Jørgensen 1997). A complex in situ technology for co-composting metal-polluted river sediments was presented by Guangwei et al. (2009), mechanically mixed with 10% in volume of wood chips, plant stems and beer-brewing waste, and inoculated with thermophilic bacteria inside a reactor capable of maintaining high temperatures. The technology made it possible to significantly stabilise Cu, Zn and Pb so as to meet the environmental quality standards for surface water in China, and to allow the reuse of the reclaimed sediments for the revegetation of the local riverbank.

The above-mentioned successful studies support co-composting as a technique capable of degrading harmful organic molecules even in contaminated mineral matrices, although not much attention has been given to the design of such co-composting processes. In this context, a co-composting study conducted by Macía et al. (2014), who adopted a co-composting approach for producing a sediment-based technosol, suitable for plant growth, as proven by the high germination index, can be considered as some pioneering work.

2.1 Effects of Additives on Composting Process and Compost Quality

The vast majority of the scientific literature focusing on co-composting reports the effects of various organic and inorganic additives on the fundamental composting processes and parameters, such as maturation, nutrient losses and heavy metal availability. For example, Lefcourt and Meisinger (2001) reported that the addition of hydrated double sulphate of K-Al or zeolite to dairy slurry at rates ranging between 0.4% and 6.25% significantly reduced ammonia volatilisation, and Venglovsky et al. (2005) reported that the addition of zeolite to pig slurry at rates of 1% and 2% reduced the compost pH value and a significantly higher concentration of water-soluble ammonia during the decomposition process, as compared to non-amended compost. The speeding up of the maturation of municipal solid waste composting process upon inoculation with fermenting bacteria and cellulolytic fungi at high microbial density ($1 \cdot 10^9$ colony-forming unit CFU/ml, 5 ml kg dry mass) was reported by Wei et al. (2007) and was ascribed to the significantly increased organic matter humification as compared to non-inoculated waste. A more effective sanitation in regard to enteric and other pathogenic bacteria was obtained by mixing composting pig manure with fly ash and lime at rates of 25%, 33% and 50%, and 4% respectively. Nishanth and Biswas (2008) reported that the addition of 4% of rock phosphate, mica and the *Aspergillus awamori*, to composting rice straw increased the P and K solubility and had higher fertilisation effects on wheat in pot experiments than a conventional NPK fertiliser. In a field trial located in the Tyrol region (Austria), Kuba et al. (2008) reported that the addition of bottom ash from wood incineration to a composting mixture of biowaste at rates of 8% or 16% improved the basic process parameters and the overall product quality, as long as low metal ashes were used, and similar results were reported by Belyaeva and Haynes (2009), although in the latter study the rate of ash addition was higher (25%). Ren et al. (2009) reported that struvite addition to composting cornstalk at rates ranging between 3.8% and 8.9% significantly reduced the total N losses and improved the maturation, compared to the unamended compost. Steiner et al. (2010) reported a faster decomposition and a significant reduction of ammonia volatilisation, and Bolan et al. (2012) reported that the addition of clay minerals and Fe-(hydro)-xides to poultry and cow manures at rates of 5% (w/w) produced an increase in the stabilisation of C attributed to its immobilisation onto the mineral phases that prevented the microbial decomposition, without a negative impact on the quality of soil amended with such compost. Concerning the dynamics of heavy metals, which generally increase their concentration due to the mass reduction during the composting process, Chen et al. (2010) reported that the addition of bamboo-derived biochar to pig manure and sawdust at rates ranging between 3% and 9% (w/w) resulted in a significant reduction of N losses and significantly lower solubility Cu and Zn. Lu et al. (2014) reported that the addition of 5% rock phosphate to pig manure and rice straw compost decreased the availability of Cu and Zn related to the increase of the compost pH value organic carbon stabilisation. Khan et al. (2014) reported that the addition of biochar addition at rates of 5% and 10% (w/w) to composting chicken manure and pine sawdust resulted in different maturation dynamics related to the increase of microbial activity, and reduced N losses from the composts by ammonia volatilisation and nitrate leaching, depending on the origin of the biochar. Similar results were obtained by Zhang and Sun (2014), who reported a faster decomposition of composting green waste co-composting with spent mushroom compost in the presence of biochar added at rates of 20% or 30%. Czekała et al. (2016) carried out 5–10% of biochar addition to mixed poultry manure/wheat straw compost and reported that biochar addition at 10% increased the temperature but shortened the length of the thermophilic phase, results that paralleled those shown by Waquas et al. (2018), who used biochar produced from lawn waste added to food waste compost at rates of 10% and 15% to increase the velocity of organic matter degradation and compost maturity, meeting the main international compost quality standard criteria. In a complex experiment involving the use of clay minerals and biochar from conifer wood as additives, used singly or in combination at rates of 25–50% and 10%, respectively, Barthod et al. (2016) reported that the addition of clay and clay/biochar mixtures to composting green waste reduced the C mineralisation, particularly in the combined clay/biochar treatments.

2.2 Effects of Additives on Nutrient Concentration and Greenhouse Gas Emissions

The increasing stringency of the atmosphere protection measures imposed on all industrial processes has led to an increase in research in compost science on potential positive impacts of various additives on the abatement of greenhouse gas emissions during the composting process. Among the earlier works on this aspect, Hao et al. (2005) reported that a treatment of livestock manure with phosphogypsum at rates of between 10% and 30% (w/w) led to a significant reduction of methane emissions, but only at the highest rate. In a study of the effects of bulking agents on the gaseous emissions of composting kitchen waste, Yang et al. (2013) showed that different bulking agents, such as cornstalks, sawdust, and spent mushroom, reduced the emissions of CH_4 and N_2O, particularly sawdust, although they were not effective in reducing NH_3 emissions. The latter study was also interesting because it attempted to calculate the C and N mass balance of the greenhouse gas emissions. Awasthi et al. (2016) reported that combinations of Ca-saturated bentonite and biochar (B) could effectively reduce the greenhouse gas emissions and nutrient losses of dewatered sewage sludge during composting. A significant reduction of greenhouse gas (CH_4, N_2O) emissions and of odour emissions (e.g. NH_3, H_2S) from composting pig manure was achieved by the addition of woody peat and Ca-superphosphate, both mixed at rates of 10%, although the use of superphosphate retarded the organic matter degradation because of an increase of the electrical conductivity values during the composting process. Effective control of emissions of NH_3 and H_2S causing odour and of various volatile compounds also impacting the atmosphere quality, such as volatile fatty acids and carbonylic compounds, was obtained by Shao et al. (2014) by the addition of rice straw at rates of 10%, 20% and 30% to composting municipal solid waste. Reduced emissions of ammonia, methane and nitrogen protoxide during the composting of duck manure was reported by Wang et al. (2014), with an additional rate of 12%, and a further reduction was obtained by the addition of earthworms to the co-composting materials 45 days after the beginning of the process. Maulini-Duran et al. (2014) investigated the effects of wood chips or chopped (0.25 × 4–10 cm) polyethylene tubes as bulking agents on the emissions of volatile organic compounds (VOCs) and greenhouse gases from a composting organic fraction of municipal solid waste in Zaragoza (Spain). Their results showed that polyethylene tube as bulking agent reduced NH_3, CH_4 and VOCs emissions more than the wood chips, the volatile organic compounds profile was dominated by terpenes, with limonene being the most abundant, and α- and β-pinene related to the wood chops bulking agent. An approach for controlling both greenhouse gas emission and the availability of heavy metals in co-composting of a 1:1 mixture of sewage sludge and wheat straw, based on the use of biochar, was presented by Awasthi et al. (2016). Their results showed that the addition of 12% of biochar and 1% of commercial lime reduced the emission of NH_3, CH_4 and N_2O, and the solubility availability of Cu, Ni, Pb and Zn. Chowdhury et al. (2014) pointed out that an optimal reduction of emissions from biochar-amended compost can be obtained when well-controlled aeration conditions can be achieved during the composting process, but this key aspect has seldom been addressed in subsequent work based on the use of biochar for minimising greenhouse gas emissions.

2.3 Main Mechanisms Identified in Compost Science and in Co-composting Approaches

Information available in the vast literature in compost science, along with practical experience gained by the broad spectrum of the composting community, has made it possible to identify some of the main mechanisms controlling the process and allows some fine-tuning to optimise the final product and minimise the environmental impact caused by the composting activities. Besides the well-established effects of temperature on the compost sanitation, the addition of sorbents, either natural or manufactured (e.g. clay minerals, zeolites, Fe-Al-(hydro)-oxides), generally reduce the ammonia volatilisation and heavy metal solubility through sorption mechanisms. Such inorganic additives can also improve the end product quality by increasing the total concentrations and solubility of key nutrients, such as K and P, once incorporated into soil. Similarly, the positive effects of biochar on compost maturation and greenhouse gas emissions can be attributed mainly to its porosity, which create additional biological space, and is also a function of its particle size distribution (Zhang and Sun 2014) in the composting mass. This can explain the lower CH_4 emissions, whereas sorption of ammonium ions that lead to reduced ammonia loss and N_2O emission. Analogously, some negative effects observed from the use of other inorganic additives, such as struvite precursors, have been related to the formation of acidic Mg-phosphate salts during the struvite crystallisation, which may acidify the composting mass, leading to lower microbial activity and reduced biodegradation, as reported by Ren et al. (2009). In all the mentioned cases, the identified mechanisms highlighted an active role of the additives in the retention of macronutrients and heavy metals with specific physico-chemical mechanisms, and not their direct role in the transformation of the organic matrix of the composting mass. To our knowledge, none of the above-mentioned studies has reported the degree of alteration of the additives at the end of the composting process regardless of the used additives, except for the dissolution of some soluble salts. In this regard, a co-composting process should make it possible to design significant physical and chemical alteration of the components in the composting mass, with their conversion into materials with innovative characteristics at the end of the process.

The co-composting of mineral environmental matrices, such as soils and sediments, mixed with organic matter, has shown potential to both accelerate the degradation of organic pollutants and to create new organo-mineral materials with texture, structure and nutrient contents that are suitable for their potential use in different agricultural sectors. In a co-composting approach to waste re-use, the organic matrix—either manure (Atagana 2004; Wan et al. 2003), sewage sludge (Ling and Isa 2006; Wan et al. 2003), green compost (Antizar-Ladislao et al. 2004) or municipal green waste (Belyaeva and Haynes 2009)—acts not only as the energy source for microorganisms (Englert et al. 1993), but also alters the reactivity of the mineral solid phases, changes the chemistry of the circulating solution, and confers a structure that conditions the movement of the liquid and gaseous phases at different scales.

Obviously, for an optimal co-composting process, the key parameters that control the microbial activity, such as the C/N ratio, the moisture content and the peak temperature and length of the thermophilic phase, must be ensured. Typical values for such parameters for an efficient composting are a C/N ratio of 25–30, as the initial optimum value for composting (Choi 1995), a moisture range of 50–70% (Liang et al. 2003) and minimum temperatures of 55–60 °C (Fan and Tafuri 1994), which favour microbial metabolism and population dynamics, effective sanitation and the degradation of eventual organic pollutants (Antizar-Ladislao et al. 2004). Temperature can be a critical aspect in the co-composting of mineral matrices, such as soils and sediments, as they do not provide metabolic energy and may represent a 'thermal sink' in the process, leading to so-called 'cold composting'. Composting at a low temperature presents drawbacks in terms of the presence of pathogens and germinating seeds, which are not killed during the process, the concentration of potentially phytotoxic compounds (Bernal et al. 2009), and a prolonged and difficult to estimate composting time to reach maturity. These drawbacks need to be overcome, mainly by arranging suitable volumes to retain heat—generally larger than those recommended for ordinary composting (UNEP-CalRecovery 2005), and preparing the co-composting materials in a small particle size so as to maximise the surface contact between the organic and mineral phases, and to speed up the composting process.

From the point of view of the main mechanisms influencing the co-composting of organic waste with mineral matrices in relatively high proportions (e.g. >30%), sorption of C and control of the pH value are expected to play major roles. In fact, microbial transformation of the organic matter mainly occurs at the liquid–solid phase interface (Bernal et al. 1998), and especially at the early stages of decomposition, sorption of low molecular weight organic matter may reduce the velocity of decomposition, leading to a cold composting process with the related outcomes listed above. Moreover, the evolution of the pH value of the co-composting mass may deviate from the typical trends due to the buffering capacity of the mineral phases, through the production of protons and other acidic chemical species, such as NH_4^+ HS^-, $H_2PO_4^-$ and low molecular weight organic acids. However, the

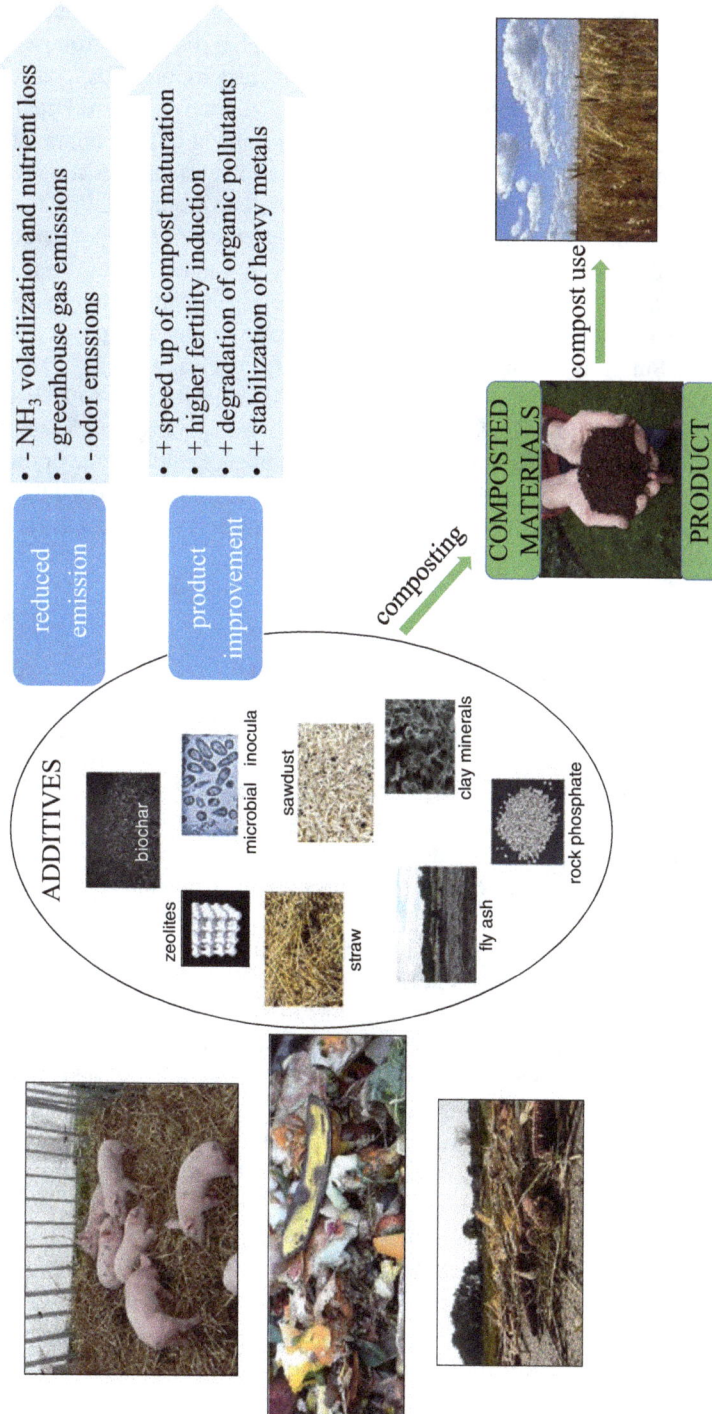

Fig. Main effects of typical additives on composting process, compost emissions and quality

optimisation of a co-composting process is feasible, as demonstrated by the work of Cai et al. (2007), who achieved a PAH removal rate higher than 94% from sewage sludge co-composted with rice straw through suitable turning methods, controlled aeration, small addition of wood chips and the inoculation of microorganisms, enzymes and growth-promoting molecules. The achievement of high temperatures is crucial for the degradation of recalcitrant and harmful substances as it increases their solubility, making them more available to metabolism (Antizar-Ladislao et al. 2004). In regard to the timing of the co-composting processes of mixed organic and mineral waste, the work of Atagana (2004) showed that the co-composting of a mineral soil with poultry manure, with the aim of soil decontamination by PAHs, required 19 months, about four to five times the typical composting time of poultry manure alone. Similarly, in the co-composting study of Huang et al. (2016) the estimated dissipation half-life of 4-nonylphenol was reduced by 3239 to 2079 days after inoculation with the degrader strain *Phanerochaete chrysosporium*, and that strain inoculation enhanced the compost maturity. Another example of the optimisation of co-composting was provided by Rekha et al. (2005), who improved the process by a liming sediment pretreatment that precipitated heavy metals in low available hydroxides.

2.4   Legislative Aspects and Implications Within the European Union

European Parliament Resolution 97/C76/01, approved on 24 February 1997, indicated (art.35) that European Union (EU) Members States should take all possible measures to guarantee the restoration of a satisfactory level at old landfill sites or other contaminated areas, suggesting diverting waste management from landfilling to the recovery and recycling of biodegradable waste and biogas production. Over the following decades, the EU established various regulatory frameworks on waste management. The EU Waste Framework Directive (Directive 2008/98/EC), and recent revision, sets the principles of waste hierarchy in order to reduce, reuse, recycle and recover, and also strategies to minimise waste disposal. The Directive requires that prevention programmes be drawn up in order to dissociate economic growth from waste environmental impacts. EU Directive 99/31/EC requires the pretreatment of biodegradable municipal waste to reach, by 2016, a progressive reduction of 35% of biowaste landfill disposal. In spite of the fact that landfilling was not encouraged, alternative waste treatment technologies, such as composting, mechanical–biological treatments, anaerobic digestion and incineration, were not specifically suggested as alternative waste management practices. The main consequence of this feature of the EU Waste Directive was that most Member States did not immediately opt for composting or biogas production, regardless of the significant benefits from these alternative forms of waste reuse. In fact, with incineration or landfilling being still the most widespread waste management options, with a biowaste recycle rate ranging between 10% and 30% (Barth et al. 2008) up to the mid-2000s. Significant diversion of waste from landfilling to composting management was achieved only when waste sorting and the removal of organic waste from landfill became mandatory in European countries, although the proportion of waste sorting is still highly variable in different areas, and composting of unsorted waste is practised (e.g. in Portugal, France and some regions of Spain).

In addition to the Landfill Directive (landfill gases endangering climate) and Waste Framework Directive, other European Economic Community programmes have fuelled the development of composting activity in Europe, among them, the EU Climate Change Programme for meeting the objectives set by the Kyoto Protocol, in terms of abatement of greenhouse gas emission from all productive sectors, and the EU Soil Protection Strategy, which highlights the need to restore the fertility of agricultural soils by a significant recycle of nutrients achievable through integrated waste management and waste reuse in agriculture. Other significant EU policies influencing the composting sector are, for example, the EU Biomass Action Plan, which aims to promote energy production from biomass; the EU target for Renewable Energies, that is 20% of energy obtained from renewable sources by 2020; the EU targets on biofuels, set to 10%; compulsory blending of biofuels by 2020; the EU programme for the development of rural areas; the Community Agricultural Practice; and the EU measures on soil conservation.

Regarding compost properties and quality, across EU Members standards that compost must meet in order to be qualified as products differ considerably. In some countries, such as Austria, France, Germany and Italy, the legislation clearly defines compost characteristics, whereas in other countries there is no harmonised legislation. Whether compost is classified as waste or not depends case by case on the decisions of the local regulatory authorities, and in some cases, it is implicitly assumed that compost is no longer a waste when it is registered as a product (Sayen and Eder 2014). Similarly, regulations and standards on compost quality are not equally established, with the exception of the limits set by Decision 2006/799/CE, as well as by the Animal By-Products Regulation. While the agronomic value (C/N ratio, minimum C content, etc.) and contaminant presence in terms of heavy metals and inert materials are usually well established in compost quality regulations, a lack of uniformity can be recognised for the direct methods used to assess pathogen presence and phytotoxicity. The lack of harmonised legislation creates uncertainty regarding waste management decisions and limits the compost productive sector. Guidelines for the use of high-quality compost in terms of material properties, plant response tests, physical contaminants and chemical properties can be found in the European Compost Network Quality Assurance Scheme (ECN-QAS) manual (https://www.compostnetwork.info/wordpress/wp-content/uploads/180711_ECN-QAS-Manual_3rd-edition_keyed-1.pdf).

In regard to the aims of this chapter, it is important to underline that, to our knowledge, the term 'co-composting' is not explicitly mentioned in any of the mentioned EU regulations, or in national or regional legislation.

3   A Proposal for an Improved Definition for Co-composting

In spite of the increase in scientific interest, co-composting is still defined as the composting of organic waste in the presence of one or more organic and inorganic 'additives', generally used at low rates, with the aim of improving specific process-related issues, such as odour emissions, quality and concentration of toxic compounds, and emission of greenhouse gases during or at the end of the composting process. Mineral, organic as well as biological additives have been shown to stimulate microbial activity, leading to an earlier start and a longer duration of the thermophilic phase as compared to regular composting. Some exceptions are considered, for example for poultry manure, which is generally recommended to be mixed at a ratio of 1:3 with lignocellulosic materials. To date, co-composting has seldom been used in the presence of 'biotic additives', such as earthworms and microbial inocula.

The term co-composting is also used when a bulking agent or wood splinters are mixed with the target organic waste to be composted. For example, composting of selected waste such as food waste bulking agents in different proportions is essential to provide a suitable structure to provide a physical habitat for the proliferation of the active microbial communities, to allow the maintenance of suitable moisture levels and to prevent anaerobic decomposition taking place. Although in some cases the bulking agents can also be used to balance the C/N ratio of the composting mass, and to supply additional available C to the microorganisms active in the organic matter decomposing, this is generally not the primary scope of its use.

Here, we propose that the term co-composting should be more properly used to refer to a designed process of composting two or more organic and inorganic matrices at various rates, to reach an intended composted product, with properties suitable for its use in agriculture and the environment, as illustrated in Fig.. Therefore, the definition of co-composting should be independent of the nature and proportion of waste in the mix, because the primary aim of the co-composting process should be the transformation of the waste from the physical, chemical and microbiological point of view to achieve the formation of a designed product.

The potential of obtaining products with tailored properties through a designed co-composting process marks the difference between the use of 'additives' that can reduce emissions, improve the composting and the product quality at the end of the process in terms of nutrients content and heavy metal solubility of the end product. Several examples of scientific reports using the term 'co-composting' are reported below. In principle, it should apply for the bulking agent, which may predominate in the composting mass in some cases, for example up to two-thirds in the composting of food waste (Eftoda and McCartney 2004). Although this may be altered during the process, bulking agents are generally used to improve the decomposition of the target organic waste. The design of co-composting should rely on the control of the main physico-chemical parameters, such as temperature and the duration of the thermophilic phase, change of the pH and salinity values, the contribution of macronutrient and micronutrients of the matrices used for the co-composting process, the concentration and speciation of organic and inorganic pollutants, changes in the

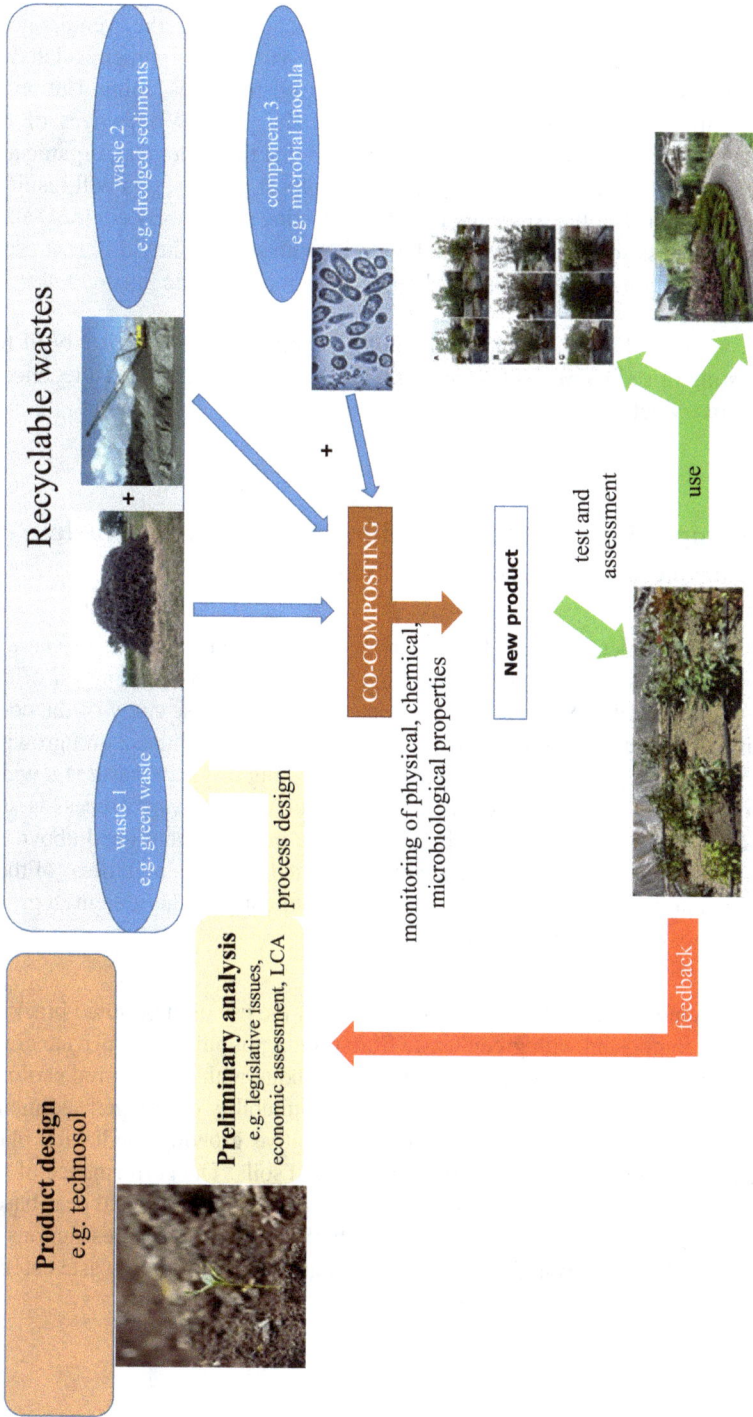

Fig. Proposed co-composting workflow from product design to compost evaluation and use

moisture content, and evolution of the microbial consortia during the entire co-composting process. Relying on this definition of co-composting, the eventual positive effects of the designed co-composting process on the environment related to greenhouse gas abatement, use of fossil fuel–derived materials, minimisation of waste landfilling, even if pursued, should not be the primary objectives of the designed process. In fact, while it has been reported that the biological, organic and mineral additives can significantly improve the compost process, this will result in the same end product of the waste under the composting process. By contrast, a co-composting process should lead to a final material with significantly different properties from the original, and with suitable properties for a specific intended use.

Below, we provide a practical example of this approach in a case study of the co-composting of sediments with green waste for their conversion into a technosol for plant nursery activities.

4   Co-composting of Dredged Sediments with Green Waste to Produce Technosols: A Pilot Study

A pilot project conducted in Italy by the authors of this chapter is presented here to show the usefulness, as well as the challenges, of the improved definition of co-composting presented in Sect.. In this section, we present a case study of the co-composting of sediments and green waste to produce fertile technosol and growing media suitable for a plant nursery. While the composting of green waste is a widespread practice, the use of dredged sediments for a co-composting process is still problematic, for several reasons, such as legislative ones, as mentioned above. In fact, in Italy, dredged sediments are currently classified as waste, regardless of their degree of contamination, and their use as soil amendment or an ingredient in growing media is not officially permitted.

In the presented case study, we demonstrated that the use of unpolluted brackish sediments co-composted with green waste from pruning of public and private green areas can produce fertile technosols and growing media, with no potential ecological risks. The case study aimed to demonstrate the suitability of dredged sediments co-composted with green waste to produce innovative growing media for plant nurseries and amendment for restoration of degraded soils. The performance of the novel sediment-based growing media was compared with that of typical peat-based growing media. The presented case study also highlights how some restrictions in the current legislation prevent the integrated management of different waste categories in co-composting treatments, regardless of their pollution levels.

4.1   Materials and Methods

Sediments (S) were dredged from the Navicelli canal (Pisa, Central Italy, 43°38′32.9″N, 10°21′19.4″E), a commercial 17 km-long channel connecting the city of Pisa to the coast that hosts numerous industrial activities. Analysis of freshly dredged sediments showed an average concentration of PAHs of 1.36 mg/kg dw, slightly exceeding the limit (1.00 mg/kg) set by Italian legislation. After dredging, sediments were allowed to dry inshore in the Navicelli area for 2 months prior to sampling, and 2 m^3 of sediments were collected from the surface layer (0–30 cm) of the sediment pile, crumbled and further air-dried. Fresh pruning waste (GW), consisting of mixed tree branches collected from public and private green areas, obtained from the waste management organisation of the city of Florence, were shredded and used within 3 days of collection. The co-composting experiment was conducted using 0.200 m^3 volume cylindrical composters and four treatments were tested: sediments only (S), GW only (GW), 1:1 w:w S:GW (SGW1:1) and 3:1 w:w S:GW (SGW3:1). All treatments were prepared in three replicates arranged in a completely randomised block design. Composters were left outdoors from June 2014 to March 2015, and tap water was added only at the beginning of the co-composting process, and on sampling occasions for the analysis of leachates. The composting materials were manually mixed after 1, 2, 3 and 6 months of composting to homogenate to allow for the optimal completion of the co-composting process. All composting materials were regularly checked for temperature and moisture content, and every 3 months subsamples were taken for the analysis of total and organic carbon and total nitrogen, and for the content of humic acids (HA) and fulvic acids (FA), to determine the humification index (HA/FA ratio). Bulk density was calculated by the weight/volume ratio of an undisturbed sample after drying at 105 °C until constant weight. Concentrations of heavy metals and PAHs were analysed by inductively coupled plasma-optical emission spectroscopy, and by extraction with acetone and hexane mixture, followed by gas chromatography–mass spectrometry (GC/MS) determination. Full details of the co-composting preparation and analytical methodologies were reported by Mattei et al. (2016). At maturity, the co-composted and parent materials were evaluated for their main physico-chemical properties, eco-toxicity, microbial activity and diversity, and fertility, and details of these aspects were reported by Mattei et al. (2017).

The results of this pilot experiment show that, notwithstanding the fact that the co-composting materials underwent an initial short and moderate thermophilic phase, resulting in a cold composting process, both the SGW3:1 and SGW1:1 products had physical and chemical properties that complied with the quality guidelines for growing media in terms of total organic C, N and humification index, pH and electrical conductivity, and bulk density values. Interestingly, the PAHs concentration in SGW3:1 and SGW1:1 were reduced by 26% and 57%, respectively, up to concentrations below 1 mg kg^{-1}, confirming that co-composting of sediment with green waste is a suitable approach for producing plant-growing

Growing media product

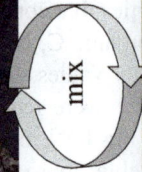

Physico-chemical parameters	S:GW 1:1	S:GW 3:1	Pruning residues	Sediments
Bulk density g cm^{-3}	0.41[c]	0.65[b]	0.19[d]	1.04[a]
pH value	8.02	8.20	8.20	8.02
EC value (mS cm^{-1})	2.92[b]	4.25[a]	0.88[c]	4.98[a]
TOC%	13.6[b]	5.52[c]	32.30[a]	1.75[d]
N%	0.42	0.25	1.15	0.17
C/N	32.0[a]	22.1[b]	28.6[b]	10.0[c]
Humic substances (total)	6588[b]	4651[c]	22191[a]	1731[c]
Humification index	H3-H6	H3-H6	H3-H6	H3-H6
PAH total (mg kg^{-1})	0.5[b]	0.5[b]	0.05[a]	0.9[a]
Phytotoxicity	<20%	<20%	<20%	<20%
Microcrustacean toxicity	<20%	<20%	<20%	<20%
Biotox	<20%	<20%	<20%	<20%

Different superscript letters in each raw indicate significant differences (P < 0.05).

Fig. Main physico-chemical properties and eco-toxicity of the plant-growing media produced by sediment co-composting with green waste

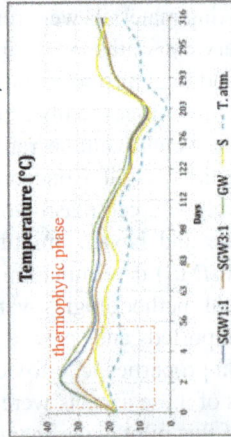

media with optimal characteristics, and also eventually degrades organic pollutants (Fig.).

Results of the eco-toxicity, microbial diversity and performance of sediment in the sediment-based growing media co-composted with the green waste as growing medium for ornamental plants showed that the co-composted materials increased the diversity of bacteria, fungi and archaea, as compared to the sediment alone, had no ecotoxicological impacts on microorganisms, micro-invertebrates and plants, evaluated with the Biotox test (Lappalainen et al. 2001), *Daphnia magna* mortality and immobilisation test (ISO 6341:1996), and the phytotoxicity test (ISO 11269-1:2012), respectively. Moreover, the co-composted material allowed an optimal growth of the ornamental plants of prime interest for the local market, *Photina* x *fraseri* and *Viburnum tinum*.

The tested treatment also increased the speed of degradation of PAHs in the slightly polluted sediments, and a decrease of salinity was a key factor in enabling an effective organic matter and PAHs degradation by the microbial community. An important aspect was that co-composting reduced the sediment bulk density, enhancing sediment aeration and permeability, to enrich the sediment with N and humic substances, and to increase microbial biodiversity.

Co-composting of dredged material with green waste proved to be a sustainable and effective treatment to convert the two waste materials into a growing medium with no eco-toxicity and high fertility.

4.2 Key Observations

A limiting factor in the use of the co-composting process as a remediation technique is the possible presence of heavy metals in raw matrices, as they tend to concentrate during the composting process, compromising or limiting the possible reuse of the finished product. In the presented case study, the concentrations of heavy metals in the co-composted sediments were all below the Italian legislation limits on growing media (Legislative Decree 75/2010) and eco-toxicity, and plant elemental concentrations did not show effects attributable to excessive heavy metals availability. However, the novel sediment-based growing media have no corresponding materials in the Italian legislation, nor in European legislation, because sediments are not currently admitted as ingredients of growing media, and therefore a comparison with the materials currently admitted in Italian legislation is not straightforward. The initial nutrients content of sediments and green waste, and the formation of a physical microstructure improving the water retention, were the main factors inducing fertility in the sediment-based growing media. These fundamental fertility factors provide an edge over the peat- and coconut fibre–based growing media, allowing plant nursery with reduced nutrient contents and alleviating water use, which has worsened in the EU, particularly in the Mediterranean area. The devised co-composting process appeared to be in line with the major EU initiatives to reduce environmental impact and soil loss, developing innovative management options for

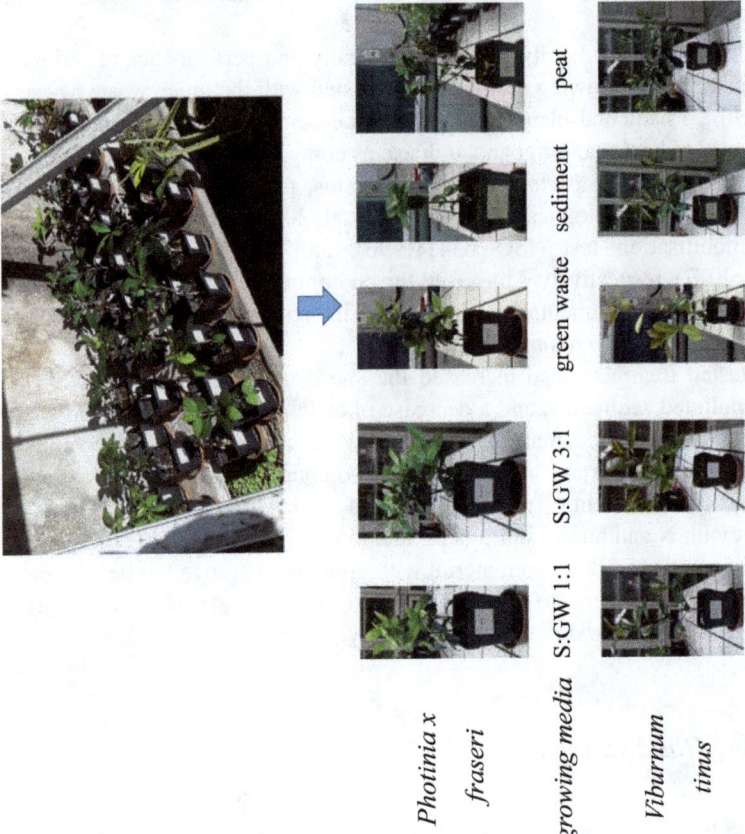

	Plant elongation (cm)				
	S:GW 1:1	S:GW 3:1	Green waste	Sediments	Peat
Viburnum tinus	3.0[a]	4.9[a]	5.3[a]	1.5[b]	1.62[b]
Photinia x fraseri	13.5[a]	9.2[b]	17.0[a]	6.3[c]	8.0[b]
	Plant biomass (g)				
Viburnum tinus	5.3[b]	7.0[a]	8.4[a]	2.3[c]	6.5[b]
Photinia x fraseri	8.7[b]	9.7[b]	9.0[b]	7.9[c]	11.9[a]

Fig. Assessment of sediment-based growing media as technical growing media for ornamental plants. (Pictures credit G. Rami)

dredged sediments and pruning residues, two relevant waste categories for various EU Countries. The results are in line with the objectives of soil protection illustrated in the EU Commission Report (COM 2012), which argues that the use of sediments co-composted with green waste should be seen as a strategy to achieve a 'land degradation-neutral world'. Implementation of this new technology at commercial scale is in line with the emission reduction targets from the agricultural sector, prevention of soil degradation and restoration of degraded soils, and the recycling of nutrients. Therefore, the results of the case study are also in line with the EU thematic Strategy on Waste Prevention and Recycling, within the frame of the above-mentioned Directive 2008/98/EC, which emphasises on the need to go for an appropriate legislative proposal based on biowaste quality parameters. Sustainable and effective management of our resources in closed loops will become the key factor for the future.

An important feature in recycling waste through co-composting is the time necessary to obtain a reclaimed and fertile product. A scaling up of the devised co-composting process in wide piles and windrows, also complemented with other low-impact technologies, such as bioaugmentation, so as to increase the temperature and prolong the thermophilic phase to reduce the maturation time and further increase the stability of the final product, is currently being tested within the ambit of the EU project AGRISED (LIFE17-ENV_IT_000269).

4.3   Legislative Issues Relating to Two Types of Waste – One Product Scenario

All EU Directives, national regulations and international conventions indicate that dredged materials should be primarily considered as natural resources that should be recycled when not flown back to their original sites. However, the current environmental legislation poses potential conflicts between EU Directives and international conventions, particularly in defining dredged materials as waste or a natural resource, with no definite solution, particularly as regards the possibility of the re-use of dredged sediments as by-products in agriculture or in composting processes. Consequently, different European countries still rely on national legislations or regulations by local authorities. While for marine sediments the London Convention has priority over EU legislation, for dredged materials from inland water bodies the decision is merely local, generating administrative stack and increased management costs. Therefore, clarification of the legislation on the possibility of using dredged sediments as components of growth media for ornamental plants, after proper testing of human and environmental safety, may lead to allowing the EU-wide use of reclaimed sediments in agriculture, particularly for the cultivation of ornamental plants and as amendment for the reclamation of degraded soils. Concerning pruning residues management, currently, the EU produces on average 13 million tonnes of wood pruning residues from various sources, with only a minor fraction used as energy biomass (GREENOVATE 2018). Although forestry and milling waste is already used for producing pellets and woodchips, pruning residues have limited use potential due to their variable and composite quality, as compared to other plant biomass (e.g. energy crops). The lack of systematic reuse of pruning residues and their remaining outside of the bioenergy market give rise to sustainability concerns. Pruning waste is treated as a plant biomass resource or as a waste. More rarely, pruning waste is shredded and left on the soil to act as a soil fertiliser or conditioner, or more rarely it is used by the farmer as firewood for self-consumption. In most cases, pruning residues are removed to prevent phytosanitary problems and have limited value as fertilisers, as they mostly present high C/N ratio values, leading to N immobilisation in soil, aggravating the low N use efficiency of most crops. There is currently no suitable large-scale value chain for pruning residues across the EU, which are commonly managed across multiple collection sites at local scale, mainly composted or landfilled. There is also uncertainty on the amount of pruning residues actually used to produce energy in the EU, because the current EUROSTAT (2014) category 'other vegetal materials and residues' includes biomass not specified elsewhere, such as straw, vegetable husks, ground nut shells, pruning brushwood, olive pomace and other waste arising from the maintenance, cropping and processing of plants. It is estimated that in Mediterranean countries orchards produce on average 2.5 tonnes of pruning residues per hectare per year, and the Czech Republic and Italy, where the HORTISED demonstration trials will be conducted, are by far the EU countries producing the highest amounts of pruning residues, with an annual production of ca. 7.8 and 5.2 million tonnes per year, respectively (GREENOVATE 2018). There is clear evidence of the impact caused by the lack of management of pruning residues in the EU, and their reuse is in line with the need to increase renewable energy to meet the 2020 climate and energy targets. The use of pruning biomass can be encouraged if their role in innovative technologies can be demonstrated, and although pruning residues are considered of low quality, they can be used to reclaim dredged sediments through a co-composting process (Mattei et al. 2016). This technological approach can stimulate the establishment of a business model that successfully takes account of the labour intensity of pruning collection and transportation, particularly in relation to the demand for growing media for plant nurseries, amendments to restore the fertility of degraded soils, and technosols for the maintenance of urban green obtained by co-composting of sediment and pruning residues. Farms have expressed an interest in changing the destination of pruning residues, particularly towards a practical and cheap recovery, suitable to allow their recycle in agriculture. The use of a significant proportion of pruning residues for the production of sediment-based growing media, soil amendment and technosols by well-designed co-composting protocols can contribute to compliance with National Renewable Energy Action Plans of the EU Member States, and also reduce the open-air burning of pruning residues, which cause environmental problems, such as the release of fumes and micro-particulate pollutants in the atmosphere.

5　Discussion

Overall, the results of the above-illustrated pilot experiment confirmed the potential reclamation of sediment through a proper design of a co-composting process with green waste, in line with the results previously obtained by Macía et al. (2014). Monitoring of PAH concentration during the co-composting process also confirmed the potential of this approach to remediate eventual high concentrations of organic pollutants, confirming previously published observations (Aparna et al. 2008; Cai et al. 2007; Huang et al. 2016; Rekha et al. 2005). The adopted co-composting approach can also contribute to a better management of the pruning residues produced by the management of green urban areas. In fact, notwithstanding the fact that the most recent EU Directive 2015/1185 (implementing Directive 2009/125/EC) and Regulation (EU) 2015/1189 also encourage the reuse of pruning residues, it is estimated that in Italy more than 80% are landfilled due to the unfavourable costs of their energy rating, that is the ratio between the pruning residues' net calorific value and the impact and costs of transportation (EBS 2019). Furthermore, pruning residues from the urban areas can be polluted by heavy metals and organic xenobiotics that prevent their direct use as soil amendment, and can result in low-grade compost.

Owing to the potential matter and energy recycling, the composting of waste materials is a non-replaceable technological approach to waste management for all human societies, independent of the degree of development (Hoornweg et al. 2013), and properly designed co-composting can enhance the sustainable management of various waste types, through the creation of a new product downstream of the process. To achieve this goal, nutrients present in the waste blends and formation of a suitable physical structure in the products can be a priori designed to fulfil specific technical demands. However, notwithstanding the ever-increasing interest in compost science, the progress in the testing of various materials and conditions, and the use of more sensitive analytical techniques to identify the main chemical mechanisms involved in compost maturation and their final properties, most of the cited works have been carried out at bench, vessel or pilot scale. The outcome of the pilot experiment reported on in this chapter will be confirmed through a scaling up at industrial level during the demonstration phase of the LIFE AGRISED project (http://www.lifeagrised.com). Overall, the obtained results indicate that a more systematic adoption of a co-composting process may provide a 'win–win' option for the integrated management of dredged sediments and pruning residues in a short supply chain, as growing media, soil amendment and technosols might be locally used, thus creating a potential development of value chain at local level.

5.1　Technical Mismatch and Legislative Gaps

Concerning the technical mismatch and sustainability issues, we notice that interesting results in regard to the control of the maturation process, nutrient retention and greenhouse gas emissions during the composting of organic waste were obtained using additives that are seldom available in different areas where composting is a practical solution for the management of organic waste. For example, are suitable amounts of clay and Fe-Al-bearing minerals commonly available in rural areas? Biochar amendment appears to be a promising additive that can effectively reduce the emissions from composting waste and stabilise the heavy metals in the final compost. Even in developed countries, is pyrolysis technology sufficiently widespread to ensure sufficient amounts of biochar for the composting industry? Moreover, because, if they are available the same additives are often requested by other industrial processes, at what cost can they be obtained by the compost industry?

In our experience, there are still limitations that apply to both the legislative and shared knowledge aspects. However, while limitations caused by poor information on new technologies can be alleviated by access to digital multimedia and dedicated communication channels between scientists and stakeholders, legislative limitations are slower to surmount, as they are related to the political agenda of different countries. For example, this is well illustrated by the case of biochar, which is officially admitted as a fertiliser or soil conditioner in some countries (e.g. Italy), but not in several others. Therefore, all the positive results related to the use of biochar can be taken advantage of only where biochar is officially permitted under the environmental and agricultural legislation. In the case of dredged sediments, while they can be employed in several civil engineering uses, their use is not permitted in agriculture, and from the legislative point of view limiting the presence of sediments in the finished product may limit their use in agriculture. Therefore, it is not clear that co-composted sediments can be used as fertilisers or soil amendments. These legislative discrepancies limit the use of innovative co-composting approaches, both at production and commercial levels.

6 Conclusions

In our opinion, the 'compost community' can substantially contribute to the minimisation of the environmental impact of the waste cycle and the maximisation of materials and energy recycling by demonstrating the possible upscaling of the most promising approaches developed at the microscale. We identified two main reasons for hindrance in our literature survey, which we termed (*i*) technical mismatch and sustainability issues, and (*ii*) legislative and knowledge gaps. We envisage even greater difficulties in the case of co-composting as we defined it in this chapter, that is the designed mixing of different types of waste to obtain new products. In this case major changes will be needed to allow the use of co-composted materials obtained by diverse sources, and clear and broad dissemination of knowledge, through the scientific and popular press, will be essential to obtain public acceptance of the innovative processes and materials.

Acknowledgement This work was supported by the EU project AGRISED (LIFE17-ENV_ IT_000269).

References

AGRISED Project, LIFE17-ENV_IT_000269. EASME, European Commission, Brussels, Belgium.

Antizar-Ladislao, B., Lopez-Real, J. M., & Beck, A. J. (2004). Bioremediation of polycyclic aromatic hydrocarbon (PAH)-contaminated waste using composting approaches. *Environmental Science and Technology, 34*, 249–289.

Aparna, C., Saritha, P., Himabindu, V., & Anjaneyulu, Y. (2008). Techniques for the evaluation of maturity for composts of industrially contaminated lake sediments. *Waste Management, 28*, 1773–1784.

Atagana, H. I. (2004). Co-composting of PAH-contaminated soil with poultry manure. *Letters in Applied Microbiology, 39*, 163–168.

Awasthi, M. K., Wang, Q., Chen, H., Wang, M., Awasthi, S. K., Ren, W., Cai, H., Li, R., & Zhang, Z. (2016). In-vessel co-composting of biosolid: Focusing on mitigation of greenhouse gases emissions and nutrients conservation. *Renewable Energy, 129*, 814–823.

Barth, J., Amlinger, F., Favoino, E., Siebert, S., Kehres, B., Gottschall, R., Bieker, M., Löbig, A., & Bidlingmaier, W. (2008). *Compost production and use in the EU*. Final report. European Commission, DG Joint Research Centre/ITPS.

Barthod, J., Rumpel, C., Paradelo, R., & Dignac, M.-F. (2016). The effects of worms, clay and biochar on CO2 emissions during production and soil application of CO-composts. *The Soil, 2*, 673–683.

Barthod, J., Rumpel, C., & Dignac, M. F. (2018). Composting with additives to improve organic amendments. A review. *Agronomy for Sustainable Development, 38*(2), 17.

Belyaeva, O. N., & Haynes, R. J. (2009). Chemical, microbial and physical properties of manufactured soils produced by co-composting municipal green waste with coal fly ash. *Bioresource Technology, 100*, 5203–5209.

Bentham, R., & McClure, N. (2003). A novel laboratory microcosm for co-composting of pentachlorophenol contaminated soil. *Compost Science and Utilization, 11*, 311–320.

Bernal, M. P., Navarro, A. F., Sanchez-Monedero, M. A., Roig, A., & Cegarra, J. (1998). Influence of sewage sludge compost stability and maturity on carbon and nitrogenmineralization in soil. *Soil Biology Biochemistry, 30*, 305–313.

Bernal, M. P., Alburquerque, J. A., & Moral, R. (2009). Composting of animal manures and chemical criteria for compost maturity assessment. A review. *Bioresource Technology, 100*, 5444–5453.

Bernstad, A., & la Cour Jansen, J. (2011). A life cycle approach to the management of household food waste – a Swedish full-scale case study. *Waste Management, 31*, 1879–1896.

Bolan, N. S., Kunhikrishnan, A., Choppala, G. K., Thangarajan, R., & Chung, J. W. (2012). Stabilization of carbon in composts and biochars in relation to carbon sequestration and soil fertility. *Science of the Total Environment, 424*, 264–270.

Büyüksönmez, F., Rynk, R., Hess, T. F., & Bechinski, E. (1999). Occurrence, degradation, and fate of pesticides during composting. I. Composting, pesticides, and pesticide degradation. *Compost Science and Utilization, 7*, 66–82.

Cai, Q. Y., Mo, C. H., Wu, Q. T., Zeng, Q. Y., Katsoyiannis, A., & Férard, J. F. (2007). Bioremediation of polycyclic aromatic hydrocarbons (PAHs)-contaminated sewage sludge by different composting processes. *Journal of Hazardous Materials, 142*, 535–542.

Chen, Y. X., Huang, X. D., Han, Z. Y., Huang, X., Hu, B., Shi, D. Z., & Wu, W. X. (2010). Effects of bamboo charcoal and bamboo vinegar on nitrogen conservation and heavy metals immobility during pig manure composting. *Chemosphere, 78*, 1177–1181.

Choi, K. (1995). Optimal operating parameters in the composting of swine manure with wastepaper. *Journal of Environmental Science and Health, 34*, 975–987.

Chowdhury, M. A., de Neergaard, A., & Jensen, L. S. (2014). Potential of aeration flow rate and bio-char addition to reduce greenhouse gas and ammonia emissions during manure composting. *Chemosphere, 97*, 16–25.

COM. (2012). *Convergence report 2012 – European Economy 3|2012*. European Commission Directorate-General for Economic and Financial Affairs.

Czekała, W., Malinska, K., Cáceres, R., Janczak, D., Dach, J., & Lewicki, A. (2016). Co-composting of poultry manure mixtures amended with biochar – The effect of biochar on temperature and C-CO2 emission. *Bioresource Technology, 200*, 921–927.

Decision 2006/799/CE. Establishing revised ecological criteria and the related assessment and verification requirements for the award of the Community eco-label to soil improvers. *Official Journal of the European Union* 24.11.2006.

Directive 2008/98/EC. On waste and repealing certain Directives. *Official Journal of the European Union* L 312/3.

Directive 2009/125/EC. Establishing a framework for the setting of eco-design requirements for energy-related products. *Official Journal of the European Union* 31.10.2009.

Directive 99/31/EC. On the landfill of waste. Council Directive *Official Journal of the European Union* L 182/1.

EBS. (2019). *Italian Association of solid biomass*. Primo Rapporto Socio Economico e Ambientale.

Eftoda, G., & McCartney, D. (2004). Determining the critical bulking agent requirement for municipal biosolids composting. *Compost Science and Utilization, 12*, 208–218.

Englert, C. J., Kenzie, E. J., & Dragun, J. (1993). Bioremediation of petroleum products in soil. In E. J. Calabrese & P. T. Kostecki (Eds.), *Principles and practices for petroleum contaminated soils* (pp. 111–130). Boca Raton, FL: Lewis Publishers.

European Parliament Resolution 97/C76/01. (11/03/1997). A Community strategy for waste management. *Official Journal C 076*, 0001–0004.

Eurostat. (2014). *Waste statistics*. Statistics explained.

Fan, C., & Tafuri, A. N. (1994). Engineering applications of biooxidation processes for treating petroleum contaminated soil. In D. L. Wise & D. J. Trantolo (Eds.), *Remediation of hazardous waste contaminated soils* (pp. 373–406). New York: Marcel Dekker.

GREENOVATE. (2018). *Mobilising pruning residues to expand Europe's biomass market*. Euro Pruning, GREENOVATE annual report.

Guangwei, Y., Hengyi, L., Tao, B., Zhong, L., Qiang, Y., & Xianqiang, S. (2009). In-situ stabilisation followed by ex-situ composting for treatment and disposal of heavy metals polluted sediments. *Journal of Environmental Science, 21*, 877–883.

Hao, X., Larney, F. J., Chang, C., Travis, G. R., Nichol, C. K., & Bremer, E. (2005). The effect of phosphogypsum on greenhouse gas emissions during cattle manure composting. *Journal of Environmental Quality, 34*, 774–781.

Hoornweg, D., Thomas, L., & Otten, L. (2000). *Composting and its applicability in developing countries*. (Urban waste management working paper series no. 8). Washington, DC: The World Bank.

Hoornweg, D., Bhada-Tata, P., & Kennedy, C. (2013). Environment: Waste production must peak this century. *Nature, 502*, 615–617.

Huang, D., Qin, X., Xu, P., Zeng, G., Peng, Z., Wang, R., Wan, J., Gong, X., & Xue, W. (2016). Composting of 4-nonylphenol-contaminated river sediment with inocula of Phanerochaete chrysosporium. *Bioresource Technology, 221*, 47–54.

ISO 11269-1:2012. Soil quality – Determination of the effects of pollutants on soil flora – part 1: method for the measurement of inhibition of root growth.

ISO 6341:1996. Water quality – determination of the inhibition of the mobility of Daphnia magna Straus (Cladocera, Crustacea) – Acute toxicity test.

Khan, N., Clark, I., Sánchez-Monedero, M. A., Shea, S., Meier, S., & Bolan, N. (2014). Maturity indices in co-composting of chicken manure and sawdust with biochar. *Bioresource Technology, 168*, 245–251.

Kuba, T., Tschöll, A., Partl, C., Meyer, K., & Insam, H. (2008). Wood ash admixture to organic wastes improves compost and its performance. *Agriculture, Ecosystems and Environment, 127*, 43–49.

Laine, M. M., & Jorgensen, K. S. (1997). Effective and safe composting of chlorophenol contaminated soil in pilot scale. *Environmental Science and Technology, 31*, 371–378.

Lappalainen, J., Juvonen, R., & Nurmi, J. (2001). Automated color correction method for Vibrio fischeri toxicity test. Comparison of standard and kinetic assays. *Chemosphere, 45*, 635–641.

Lefcourt, A. M., & Meisinger, J. J. (2001). Effect of adding alum or zeolite to dairy slurry on ammonia volatilization and chemical composition. *Journal of Dairy Science, 84*, 1814–1821.

Legislative Decree 75/2010. Riordino e revisione della disciplina in materia di fertilizzanti. *Official Journal of Laws and Decrees of the Italia Republic* (GU Serie Generale – Suppl. Ordinario n. 106) n.121.

Liang, C., Das, K. C., & McClendon, R. W. (2003). The influence of temperature and moisture contents regimes on the aerobic microbial activity of a biosolids composting blend. *Bioresource Technology, 86*, 131–137.

Ling, C. C., & Isa, M. H. (2006). Bioremediation of oil sludge contaminated soil by co-composting with sewage sludge. *Journal of Scientific & Industrial Research, 65*, 364–369.

Lu, D., Wang, L., Yan, B., Ou, Y., Guan, J., Bian, Y., & Zhang, Y. (2014). Speciation of Cu and Zn during composting of pig manure amended with rock phosphate. *Waste Management, 34*, 1529–1536.

Macía, P., Fernández-Costas, C., Rodríguez, E., Sieiro, P., Pazos, M., & Sanromán, M. A. (2014). Technosols as a novel valorization strategy for an ecological management of dredged marine sediments. *Ecological Engineering, 67*, 182–189.

Mattei, P., Cincinelli, A., Martellini, T., Natalini, R., Pascale, E., & Renella, G. (2016). Reclamation of river dredged sediments polluted by PAHs by co-composting with green waste. *Science of the Total Environment*, 567–574.

Mattei, P., Roberta Pastorelli, R., Rami, G., Mocali, S., Giagnoni, L., Gonnelli, C., & Renella, G. (2017). Evaluation of dredged sediment co-composted with green waste. *Journal of Hazardous Materials, 333*, 144–153.

Maulini-Duran, C., Artola, A., Font, X., & Sánchez, A. (2014). Gaseous emissions in municipal wastes composting: Effect of the bulking agent. *Bioresource Technology, 172*, 260–268.

Namkoong, W., Hwang, E. Y., Parka, J. S., & Choi, J. Y. (2001). Bioremediation of diesel-contaminated soil with composting. *Environmental Pollution, 119*, 23–31.

Nishanth, D., & Biswas, D. R. (2008). Kinetics of phosphorus and potassium release from rock phosphate and waste mica enriched compost and their effect on yield and nutrient uptake by wheat (Triticum aestivum). *Bioresource Technology, 99*, 3342–3353.

Regulation (EU) 2015/1189. Commission Regulation (EU) 2015/1185 With regard to ecodesign requirements for solid fuel local space heaters. *Official Journal of the European Union* L 193/1.

Rekha, P., Suman Raj, D. S., Aparna, C., Hima Bindu, V., & Anjaneyulu, Y. (2005). Bioremediation of contaminated lake sediments and evaluation of maturity indices as indicators of compost stability. *International Journal of Environmental Research and Public Health*, 251–262.

Ren, L., Schuchardt, F., Shen, Y., Li, G., & Li, C. (2009). Impact of struvite crystallization on nitrogen losses during composting of pig manure and cornstalk. *Waste Management, 30*(5), 885–892.

Sayen, H., & Eder, P. (2014). *End-of-waste criteria for biodegradable waste subjected to biological treatment (compost & digestate): Technical proposals* (p. 308). Luxembourg: European Commission EUR 26425 – Joint Research Centre – Institute for Prospective Technological Studies, Publications Office of the European Union.

Shao, L. M., Zhang, C. Y., Wu, D., Lü, F., Li, T. S., & He, P. J. (2014). Effects of bulking agent addition on odorous compounds emissions during composting of OFMSW. *Waste Management, 34*, 1381–1390.

Steiner, C., Das, K. C., Melear, N., & Lakly, D. (2010). Reducing nitrogen loss during poultry litter composting using biochar. *Journal of Environmental Quality, 39*, 1236–1242.

UNEP-CalRecovery. (2005). *Solid waste management*. Report to Division of Technology, Industry, and Economics, International Environmental Technology Centre, United Nations Environmental Programme, Osaka, Japan- CalRecovery, Inc., California, USA, p. 525.

Venglovsky, J., Sasakova, N., Vargova, M., Pacajova, Z., Placha, I., Petrovsky, M., & Harichova, D. (2005). Evolution of temperature and chemical parameters during composting of the pig slurry solid fraction amended with natural zeolite. *Bioresource Technology, 96*, 181–189.

Wan, C. K., Wong, J. W. C., Fang, M., & Ye, D. Y. (2003). Effects of organic wastes amendments on degradation of PHAs using thermophilic composting. *Environmental Technology, 24*, 23–30.

Wang, J., Hu, Z., Xu, X., Jiang, X., Zheng, B., Liu, X., Pan, X., & Kardol, P. (2014). Emissions of ammonia and greenhouse gases during combined precomposting and vermicomposting of duck manure. *Waste Management, 34*, 1546–1552.

Waquas, M., Nizami, A. S., Aburiazaiza, A. S., Barakat, M. A., Ismail, I. M. I., & Rashid, M. I. (2018). Optimization of food waste compost with the use of biochar. *Journal of Environmental Management, 216*, 70–81.

Wei, Z., Xi, B., Zhao, Y., Wang, S., Liu, H., & Jiang, Y. (2007). Effect of inoculating microbes in municipal solid waste composting on characteristics of humic acid. *Chemosphere, 68*, 368–374.

Yang, F., Xue, G., Qing, L., Yang, Y., & Luo, W. H. (2013). Effect of bulking agents on maturity and gaseous emissions during kitchen waste composting. *Chemosphere, 93*, 1393–1399.

Zhang, L., & Sun, X. (2014). Changes in physical, chemical, and microbiological properties during the two-stage co-composting of green waste with spent mushroom compost and biochar. *Bioresource Technology, 171*, 274–284.

Urban Waste as a Resource: The Case of the Utilisation of Organic Waste to Improve Agriculture Productivity Project in Accra, Ghana

Dzidzo Yirenya-Tawiah, Ted Annang, Benjamin Dankyira Ofori,
Benedicta Yayra Fosu-Mensah, Elaine Tweneboah- Lawson, Richard Yeboah,
Kwaku Owusu-Afriyie, Benjamin Abudey, Ted Annan, Cecilia Datsa,
and Christopher Gordon

Abstract Poor municipal solid waste management continues to be a daunting issue for municipal authorities in Ghana. Major cities generate 2000 tonnes of mixed municipal waste per day, of which about 80% is collected and disposed of at open dump sites and/or at the limited number of landfills available. About 60% of this waste is organic. The Utilization of Organic Waste to Improve Agricultural Productivity (UOWIAP) project sought to co-create knowledge through a private-public engagement for the development of organic waste value chain opportunities to sustainably manage municipal organic waste and, at the same time, improve urban farm soils and increase food productivity in the Ga-West Municipal Assembly in the Greater Accra Region of Ghana. Through the project, identified key stake-holders in the waste and agricultural sectors, such as market traders, informal waste collectors, unemployed persons, farmers, landscapers, media, agricultural extension

D. Yirenya-Tawiah (✉) · T. Annang · B. D. Ofori · B. Y. Fosu-Mensah
E. T. Lawson · C. Datsa · C. Gordon
Institute for Environment and Sanitation Studies, University of Ghana, Accra, Ghana
e-mail: dzidzoy@staff.ug.edu.gh; tyannang@ug.edu.gh; bdofori@staff.ug.edu.gh;
yayramensah@staff.ug.edu.gh; elaine_t@staff.ug.edu.gh; cdatsa@staff.ug.edu.gh;
cgordon@staff.ug.edu.gh

R. Yeboah · K. Owusu-Afriyie
Management for Development FoundationTraining and Consultancy, Accra, Ghana
e-mail: rye@mdf.nl; koa@mdf.nl

B. Abudey
Ga-West Municipal Assembly, Local Government Office, Accra, Ghana

T. Annan
Ministry of Food and Agriculture, Accra, Ghana

officers, Municipal Assembly officers and the general public, were engaged and made aware of sustainable organic waste management processes, including organic waste segregation from source, collection and compost production. Four formal markets were selected for the piloting of organic waste segregation from source. Interested persons were trained in organic waste collection, compost production and entrepreneurship. The lessons learned draw attention to the need for a massive effort to generate demand for compost use as this will invariably drive removal of organic waste from the unsorted waste stream.

Keywords Market waste · Waste to resource · Waste segregation · Community project · Compost · Pollution prevention

1   Introduction

The challenge of municipal solid waste management continues to be a daunting issue and is of global concern. Increasing population and urbanisation, and the accompanying high rate of waste generation, have engendered serious concerns regarding the attainment of the Sustainable Development Goals. According to the World Bank, the global urban population in 2012 was about 3 billion, generating about 1.3 billion tonnes of solid waste annually (Hoornweg and Bhada-Tata 2012). Waste generation in sub-Saharan Africa was estimated to be approximately 62 million tonnes annually, reflecting an average of 0.65 kg/capita/day (Hoornweg and Bhada-Tata 2012). With a projected urban population of about 4.3 billion by 2025, municipal solid waste generation is estimated to exceed 2.2 billion tonnes (Hoornweg and Bhada-Tata 2012). The estimated projections raise concerns about the anticipated exponential increase in the volume of waste that will be produced in the near future.

The end-of-pipe approach to waste management in the twentieth century can be blamed for the outcomes and the impact of waste management observed today. The release of greenhouse gas emissions from waste landfills and its effect on climate change, the burden of gastro-intestinal diseases, the indiscriminate dumping of waste especially in developing countries and the heavy financial cost of waste management have been overwhelming (Starovoytova 2018). Faced with this daunting situation, global efforts have been directed towards reversing this trend and finding sustainable approaches to waste management.

Recent global trends in city waste management have seen a focus on managing waste in a socially and environmentally acceptable manner (Vergara and Tchobanoglous 2012), to promote public health and enhance resource use efficiency. Various frameworks and concepts have been developed to guide the sustainable management of waste. Examples include the waste hierarchy framework which outlines the most preferred waste management strategies, such as waste recovery, reduction, reuse and prevention, towards the apex of the pyramid, and the unwanted methods, such as landfilling, at the base of the pyramid. Zero waste and circular economy concepts are nowadays driving the international waste management approaches.

Waste governance is also becoming regionalised (Vergara and Tchobanoglous 2012). In developed nations, where citizens produce far more waste, the waste generated is often managed formally at a municipal or regional scale. In developing nations, where citizens produce less waste, most of which is organic, a combination of formal and informal actors is often involved in the management of waste. Thus, effective waste management strategies should vary depending on the local waste characteristics, which vary with cultural, climatic and socio-economic variables and institutional capacity.

In Ghana, solid waste generation currently ranges between 0.2 and 0.8 kg/person/day, with an estimated volume of 13,500 tonnes of solid waste being produced daily nationwide (Miezah et al. 2015). The waste composition of Ghana is predominantly organic (60%), followed by plastics (14%), paper (5%), metals (3%) and glass (3%). Major cities in Ghana generate 2000 tonnes of mixed municipal waste per day, of which 80% is collected (Cofie et al. 2009). Collected waste is disposed of either by open dumping or in the few available landfills. Ghana currently has only five engineered landfills; however, most of them are dysfunctional. Accra, for example, has no landfill site; therefore, most of the waste collected from the city is taken to Kpone in Tema, a city 24 km from Accra.

In line with recent global trends, Ghana is working on promoting waste recovery for reuse and recycling. The informal sector is actively involved in this approach to waste management. Waste pickers often collect plastics, metals and cans from the municipal waste stream and sell them to recycling companies or for reuse purposes. In Ghana, there is the potential of harnessing large volumes of organics from the municipal waste stream: compost production can divert organic waste away from landfill. Compost production can create employment and revenue opportunities for the unemployed population, farmers and waste pickers who may take part in the recovery, processing and utilisation (Gabbay 2010). Nevertheless, composting as a waste management tool is only feasible in cases where there is a strong and stable demand for compost at the local level.

The Institute for Environment and Sanitation Studies at the University of Ghana, the MDF West Africa Training and Consultancy, the Ministry of Food and Agriculture (MOFA) and the Ga-West Municipal Assembly partnered, with the funding support of the Ministry of Foreign Affairs of the Netherlands, through the Netherlands Organisation for Scientific Research (NWO) and the Food and Business Applied Research Fund, to co-create knowledge and demonstrate organic waste valorisation in an urbanised location in the city of Accra. The overall goal of the UOWIAP project was to co-create knowledge in order to promote the development of organic value chain opportunities, to enhance the food security and livelihoods of the peri-urban poor in the Ga-West Municipality. The project's objectives were to:

- Co-create knowledge for sustainable organic waste management in the municipality
- Develop organic waste value chain opportunities by engaging unemployed youth and urban farmers in the Ga-West Municipality
- Divert organic waste from the market waste stream for compost production for the local market

- Promote the practice of waste segregation amongst market traders in project markets
- Foster the use of compost amongst urban farmers
- Develop outreach programmes to communities to increase awareness of waste being a resource
- Create awareness amongst the general public on the value of compost-grown foods and their impact on the quality of life and well-being

2 The Context

The project was implemented in the Ga-West Municipality, which is located within latitudes 50°48' north and 5°39' north and longitudes 0°12' west and 0°22' west and which covers a land surface area of 299.578 km² (Ghana Statistical Service 2014). The municipality is bordered to the north by the Akwapim South Municipality and to the south by the Ga-South Municipality. It is bordered to the north-south by the Ga-Central Municipality and to the east by the Accra Metropolitan Area and the Ga-East Municipality.

The Ga-West Municipality has a population of about 219,800 inhabitants, made up of 51% females and 49% males. About 63% of the population is between the ages of 15 and 60 (Ghana Statistical Service 2014). The overspill of the urban

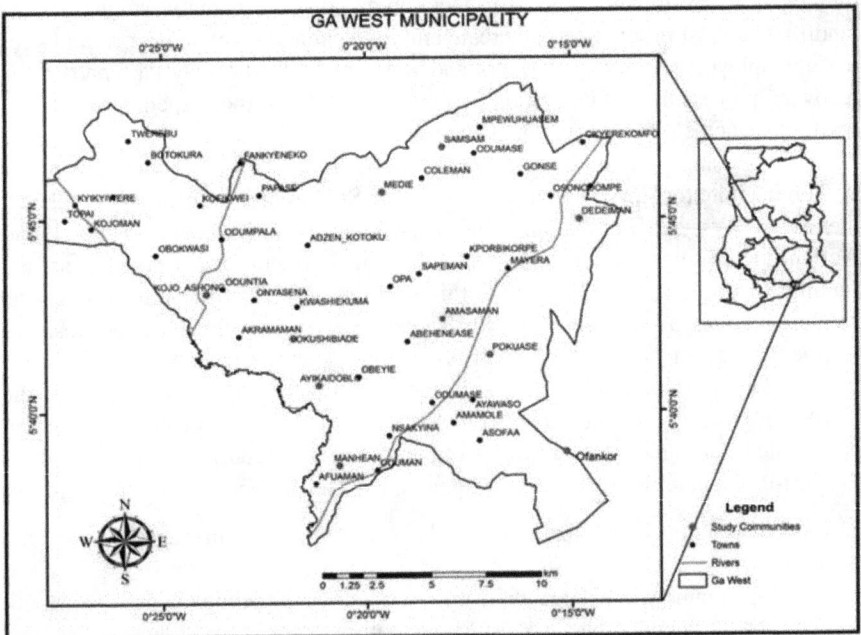

Fig. Map of the Ga-West Municipality showing the study areas

population into the peri-urban Greater Accra Metropolis has created severe challenges in waste management, urban agricultural productivity, land management and general socio-economic development. Population inflows into the study area, the Ga-West Municipality, have brought the population density to the unprecedented level of 711 persons/km^2, compared to the population density within the city of Accra of 90/km^2 and the national density of 79/km^2 (Ga-West Municipal Assembly 2010). A significant number of the population live in scattered rural settlements covering about 60% of the land area of 462 km^2. The municipality is noted for intense agricultural activities. Today, cultivated farmlands and rural neighbourhoods are gradually becoming small towns, with increasing commercial activities.

The municipality is faced with many challenges. These include farmers losing their farmlands to commercial and residential developers and sand winners. The situation has resulted in a reduced land area for farming, loss of soil fertility because of continuous cropping and increased inorganic fertiliser usage, leading to low agricultural productivity. The municipality also faces waste management challenges where volumes of uncollected municipal waste, often left in the open, are burnt amongst open-air residences or clog up natural and artificial waterways.

Agriculture remains the main occupation in the municipality, supporting around 55% of the economically active population. Within this context of declining agricultural productivity, loss of livelihoods and poor environmental sanitation in the Ga-West Municipality, there are existing opportunities to:

- Increase livelihood opportunities by reclaiming sand-won land and creating waste management jobs
- Boost soil productivity by training farmers in composting and sustainable, organic soil management
- Manage environmental sanitation by promoting the valorisation of organic waste
- Improve public health through community outreach on understanding the benefits of organic produce

2.1 Strategic Focus and Methods Used

The project addressed four strategic areas:

(i) Stakeholder identification and engagement for knowledge co-creation with project partners
(ii) Training and skills development for organic waste value chain development
(iii) Business model development
(iv) Project awareness creation

Prior to project implementation, a baseline study was conducted to ascertain the prevailing situation regarding the profile of vulnerable groups in the municipality, waste management practices in the local markets and compost production and use in the municipality amongst farmers. Community fora were held at different levels within the municipality to create awareness of the project, to solicit community support and to seek ideas for sustainable strategies needed to facilitate its implementation and the recruitment of unemployed youth for skills training. Media coverage (television, radio and news print) of community engagement activities was also used to create awareness of the project amongst the public. Findings from the baseline study informed the project implementation direction, subjects and study markets and the selection of farms. The project implementation plan was structured as follows:

1. Stakeholder workshops and meetings:

   (a) Project implementation planning.
   (b) Assess and receive input during various stages of project, including results dissemination.
   (c) Knowledge co-creation sections.

2. Waste collection and compost production training:

   (a) Develop curriculum and content for compost production and entrepreneurship training.
   (b) Recruit and train the unemployed in organic waste collection and compost production.
   (c) Conduct compost production training for farmers' groups and agricultural extension officers.
   (d) Closely monitor farmers' progress and obstacles.

3. Entrepreneurship and business creation support infrastructure
4. Youth composting scheme pilot:

   (a) Acquire land to use as a composting site.
   (b) Collaborate with partners to obtain the materials and infrastructure necessary to pilot the waste collection and composting.
   (c) Train youth on waste separation and how to educate their customers, on collection and on processing/reselling.
   (d) Build capacity for groups to manage and maintain composting facilities, produce quality compost and appropriately price and market it.

5. Community outreach:

   (a) Conduct advocacy campaigns for waste separation from source and organic fertiliser usage.
   (b) Develop strategies to involve local media, markets and schools in the promotion of organic foods (e.g. posters, leaflets and community FM stations).

6. Monitoring and evaluation of youth composting and farmer training pilots

The project was implemented for 3 years. The first year was used for sensitisation, study subject engagements, knowledge co-creation activities, developing methodologies and training, pilot testing and improvement of methodologies. In the second year training continued and field experimentation was conducted to demonstrate the value of crops and organic waste opportunities. Feedback loops were also implemented to improve the methodologies employed. The third year was for the establishment of a small-scale business venture by the trained groups. The project was structured as a trans-disciplinary pilot study.

2.2 Project Actors

The main project actors were all located within the municipality. Different key stakeholders were involved based on the different roles they play in the development of the waste value chain. These included the governmental agency staff, such as the Environmental Health Officers, Planning Officer and Youth Employment Officer of the Municipal Assembly, the Director of the MOFA and his extension officers. Others included local community leaders, community members, market queens and leaders and traders, waste pickers and caretakers of skips for waste collection, farmer groups and individual farmers, youth groups, schools and local radio stations. External stakeholders comprised the project team from the University of Ghana and MDF West Africa Training and Consultancy, the media and the general public.

3 Data Collection and Analyses

The study design was cross-sectional and it employed the use of both quantitative and qualitative methodologies. Various methodologies were adopted and tailored for the specific components of the project. For example, in order to assess the food-insecure in the study communities, participatory maps, focus group discussions, semi-structured individual interviews and key informant interviews were used.

A two-stage sampling technique was used. The first stage involved a purposive sampling of eight communities after reconnaissance visits to the district: this is where the researchers chose specific people who had knowledge of waste separation and composting to participate in the project. Two focus group discussions were organised in each of the selected communities, separate ones for men and women. Participants in focus group discussions were also selected voluntarily to obtain a good representation of key members within the communities. The results of the focus group discussions were important in developing the questionnaire for the individual interviews. Transect walks were organised with some community members to identify key locations within the communities, as well as settlement patterns. The second stage of data collection considered a random sampling to select the households for the interviews. According to the United Nations Food and Agriculture Organization (FAO) (2004, p. 50), households normally "comprise individuals who live in the same dwelling and who have common arrangements for basic domestic and/or reproductive activities such as cooking and eating". In each household, the head of the household and one other adult, in most cases the wife, were interviewed.

For market surveys, three formal markets (designated sites and markets approved by the Municipal Assembly) in the municipality and one other market in a neighbouring municipality market were engaged. Visits were conducted to the neighbouring markets to identify market leaders in order to engage and observe first-hand the state of waste management in the selected site. A survey was also conducted amongst the traders to solicit their views for, and inputs to, the project. Questionnaires were also developed to assess the waste generated from the markets and the willingness of traders to engage in waste segregation. The questionnaires consisted of both open-ended questions and closed-ended questions and centred on demographics and socio-economic characteristics, solid waste management practices and willingness to segregate and pay for segregated waste to be collected.

Another survey was also carried out to assess the profile of the farmers in the municipality. Data was collected by conducting a face-to-face interview with farmers. A pre-tested structured questionnaire and an unstructured questionnaire were developed and administered to 155 farmers (four zones), to collect data on household characteristic, types of crops cultivated in the district, land size, crop yield and source of capital for farming. The sampling method used was purposive random sampling: the researchers identified specific farmers who were willing to answer the questions.

All data was entered into SPSS version 20 (SPSS, Chicago, IL, USA) and analysed using descriptive and logistic regression statistics. Actual counts and percentages were used to describe the characteristics of the sample. Inferential statistics were set at $P < 0.05$.

4   Baseline Situation

4.1   Profile of Vulnerable Groups in the Municipality

The profiling of vulnerable groups was carried out to help target groups who could be engaged in organic waste business. The study therefore sought to understand the characteristics of the vulnerable groups in the municipality. Perceived vulnerability and poverty are context-specific. In some communities, participants indicated that a poor person is one "who cannot afford three square meals a day". In some areas a person was presumed to be poor if they do not own land. Especially in rural areas, a person was deemed to be poor if they do not own their house. According to some respondents, some people cannot afford three square meals a day and cannot care for their dependents and therefore are classified as poor people. These include the unemployed, lazy, disabled, weak or aged people.

Men were perceived as being more vulnerable to poverty. Almost 48% of study subjects believed that men form the majority of poor people, as compared to women, because men often have fewer livelihood options. In one of the study communities, everyone in the community, including the chief, classified themselves as poor. According to the participants, both the youth and the aged face similar problems because they have no lands on which to farm. Results from the interviews showed that the majority of the poor are identified to be mainly farm labourers, unemployed and beggars.

4.2   Waste Management Practices in Local Markets

A total of 108 questionnaires were administered in the three selected markets. The respondents included 88 females and 20 males. The majority of them were within the age group of 21–40 years. Most of the respondents were married. A substantial number of them (20%) had no formal education. Out of those with some form of education, the majority of them (36%) had attained middle school/junior high school level. Most of them were traders but a few (12%) were artisans who engaged in vocations such as hairdressing and sewing.

Regarding their waste management practices, the study revealed that over 90% of the interviewees dump the waste they generate into a communal skip in the markets. A few of them (7.5%) burn the waste and (0.9%) dump it indiscriminately. Those who dump at the communal skip pay to do so. The survey also revealed that only 20.8% of market women segregate their waste. Segregated wastes – mainly cassava and plantain peels – are given out for animal feed. The majority of the market traders also indicated that they have no education in waste management. Key stakeholders who were identified as being involved in market waste management included the skip supervisors, waste pickers, private service providers and market traders.

All market traders dump their mixed waste in skips provided by the Municipal Assembly. However, in two of the markets, the traders had engaged individuals who help to maintain the dump site. There was no waste dump site in the third market; as a result the traders use a communal skip placed a few metres away from the market. There was a female leading the waste dump site management in one of the markets. This woman engages young men, who support her in maintaining the dump site.

Engagement with the dumpsite managers revealed their willingness to support the project with any waste segregation effort, such as maintenance of bins for only segregated waste. However, their major challenge was the difficulty in controlling the type of waste dumped as many of the neighbouring households do not have toilets. After several consultations with stakeholders, it came out clearly that to achieve success in the area of segregated waste, the market women need to find their own designated location to position bins solely for the collection of organic waste. This site should not be close to the official dumpsite, so the market women can manage the waste collection process.

4.3 Farmers' Profile

From the survey data, the majority (63.9%) of the farmers interviewed were males, with only 36.1% being female. The results show that the majority of the farmers were within the 31–60 years age group. The 51–60 years age group (30%) formed the majority, followed by the 41–50 years age group (24%) and the 31–40 years age group (23%), with farmers above 70 years forming the smallest age group, at 2%. The majority of the farmers had no formal education (34.2%), with most of them having a basic school level of education. Only 15% of them had a secondary education. Larger part of farmers within the municipality rent (32%) and lease (32%) their farmlands. Another 31% of respondents indicated that they inherited their land; 4% reported that they clear freely available lands, whilst only 1% indicated that they bought their land strictly for farming. When questions about the number of years allowed to cultivate on rented lands were posed, the majority (47%) of respondents indicated that they are allowed to cultivate on the land for more than 5 years, whilst 35% indicated that they are allowed to cultivate for 1–2 years; the smallest group (18%) said that they are allowed to cultivate for between 3 and 4 years.

Generally, farmlands are small and fragmented. Cumulatively, the majority (35%) of the farmers have between 4000 and 8100 m² of farmlands, followed by farm size of 12,000 and 16,200 m² (29%). Only 6% of the respondents indicated that they have farms sized <4000 m² .

Fig. Source of farmland acquisition

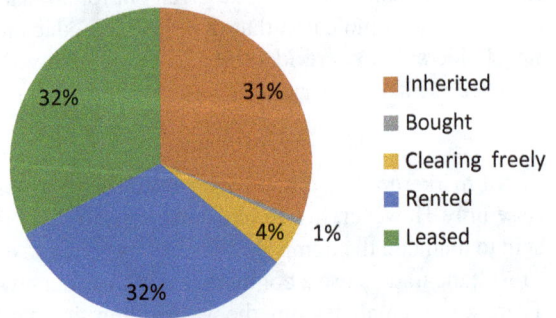

Fig. Proportion of farmers with various farm sizes

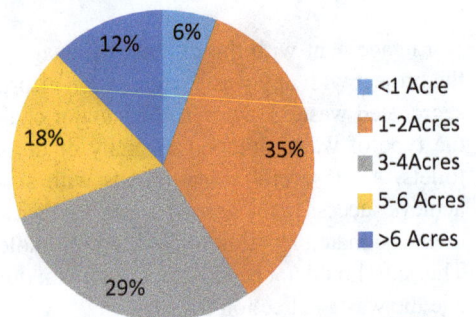

UOWIAP Field Work Progress Report 2016

Most farmers (36.1%) cultivate at least two crops in a year, whilst 33% indicated that they cultivate three crops per season. Only 2.5% of farmers cultivate more than five different crops per season, whilst 5.1% and 22.5% produce one and four crops per season, respectively. From the survey, the major crops/vegetables cultivated in the area are maize (*Zea mays*), pineapple (*Ananas comosus*), okra (*Abelmoschus esculentus*), cassava (*Manihot esculenta*), tomatoes (*Solanum lycopersicum*), pepper (*Capsicum*), garden eggs (*Solanum melongena*), lettuce (*Lactuca sativa*) and cabbage (*Brassica oleracea* var. *capitata*).

Farmers' Knowledge and Utilisation of Compost

When the question of the use of compost was posed to farmers, it was found that only 12% of them acknowledged its use. More than a half of the farmers (63%) indicated that they do not use any form of organic fertiliser on their farms, whilst 37.4% indicated that they apply some form of organic manure, which is mainly animal droppings. Danso et al. (2006) earlier reported that very few urban farmers in Ghana make use of compost.

Surprisingly, the municipality is home to one of the largest compost manufacturing companies in Accra, yet its patronage by the farmers remains low. Composting at the small-scale level is non-existent in the municipality. The willingness of farmers to use compost observed in this study reveals that the Ga-West Municipality has the potential to be a major compost hub, in both production and utilisation. Many of the farmers interviewed complained about the loss of soil fertility and acknowledged the fact that compost is likely to improve their soil quality. However, when presented with a maximum price scenario, this study found a variation in the willingness to buy. Since most farmers are producing on less than 18,000m² farm plots, it is not surprising that they expressed an unwillingness to pay anything above a GH¢ 40.00 (about 8 USD – at an exchange rate of 1 USD to 4.9 GH¢) for 50 kg. Doing so would increase their cost of farming and, inevitably, produce a perceived reduction in their profit margins. The lack of experience with compost or previous exposure to compost use could also account for the reservation in committing much money to it, because of the fear of not getting the expected income from using compost. Cofie et al. (2009) also identified the lack of a ready market as one of the major constraints to the successful operation of composting projects.

4.4   Public Views on Compost-Grown Foods

The success of organic agriculture, and by extension compost use, depends to a large extent on consumer demand (Bonti-Ankomah and Yiridoe 2006). This study revealed that all consumers were willing to buy compost-grown crops. These decisions were based on the assumption that organic crops taste better and are preferred, health-wise, to those grown with inorganic fertiliser. This supports the conclusion of Bonti-Ankomah and Yiridoe (2006) that consumer preference for organic crops is influenced by inherent product characteristics, such as the taste and appearance of the product.

About 73% of the youth in the municipality were willing to go into composting as a source of income. Analysis revealed that the need for income and composting knowledge were factors that significantly influenced the decision to go into composting. It is expected that the monetary benefit of composting would be a motivating factor that would draw people to composting, including those already engaged in some other forms of economic activity. It is therefore reasonable for the aforementioned percentage of the respondents to express a willingness to go into composting as an economic venture, even when 49% were already economically active.

Knowledge of the composting process, as a factor influencing the decision to go into composting, is significant, considering the observation that there was no previous knowledge of composting amongst the participants of the study. This confirms the study by Nartey (2013) that people are willing to go into composting when they are equipped with the necessary training. The labour required during composting, the duration of composting and the space for composting are identified as factors that are likely to deter youth from going into composting. Since income is a significant factor in the decision to go into composting, the lengthy composting period and the labour-intensive nature of the process will deter people from engaging in it. Also, space for composting is a likely deterring factor due to the likely hostile reaction of neighbours to the idea of waste being accumulated in their neighbourhood and the contentious issues normally related to land.

4.5 Exploration of Business Models

The project partner, MDF West Africa, conceptualised four business models that could potentially be implemented by the project. These models are presented in Fig.. Box 6.1 explains the money-making opportunities for the four models. In

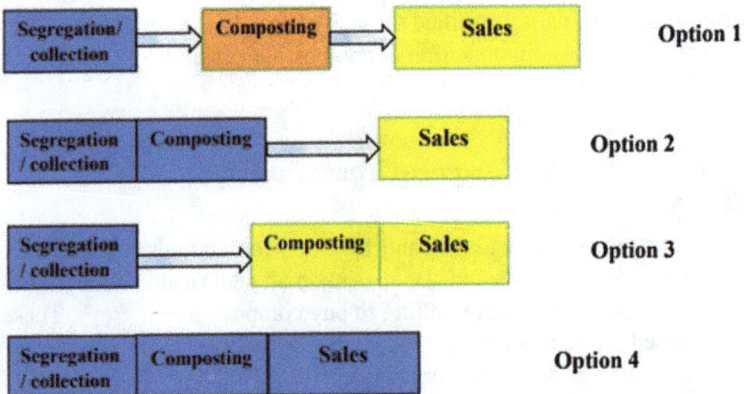

Fig. Proposed business models for organic waste valorisation for unemployed youth

Box Business model options
- Option 1: Segregation and collection give enough money if there is a big plant.
- Option 2: Only if one has enough money and the markets around are well organised.
- Option 3: One needs a good network in distribution.
- Option 4: Is only possible if one has big investment (money) and good market.

general, the four models differed based on the capacity of the entrepreneur (in terms of planning, organisation and funding). Option 1 explored the situation where the trained personnel could focus on promoting and collecting only segregated waste from the market and delivering to a producing plant for a fee. Whilst this was a potential option for the trainees, since there was an already existing plant, this option could not be implemented because the compost plant already had its staff and system of operation. Option 4 was also not an option to consider since it required substantial investment. Options 2 and 3 were therefore potential options that were tested with project support.

5   Observations from Project Implementation

5.1   Knowledge Co-creation

Knowledge co-creation was a key strategy employed to gain community support for the project, to facilitate implementation and to promote sustainability. The strategy was developed by the relevant actors. For instance, dealings with market traders were led by the traders' leadership, who engaged their colleagues, Municipal Assembly staff, waste pickers in the markets and waste collectors and with the project team devised their own workable strategies. Similarly, farmer engagement was based on open fora and several group meetings held with farmers and extension services of MOFA. This promoted ownership of the project and enhanced implementation process management and the acceptability of the project.

5.2   Youth Engagement and Skill Training

Youth engagement was quite challenging, even with the support of the Youth Employment Officers at the Municipal Assembly and radio and community fora. The youth were generally not interested in the waste business because it was viewed as a lower-level, dirty job, with minimum returns. They complained about stigmatisation and that the general public did not respect persons engaged in waste business, especially waste pickers.

To encourage participation, the criteria for selection were enhanced to include interested persons rather than a focus on youth. Some financial incentives were given to support trainees' travel to the training centre. In this way, the project attracted diverse groups of persons, which included some unemployed youth, retirees and persons seeking to learn skills for self-reliance. It was also interesting to observe that about 60% of those that enrolled to participate in the 3-month programme were women. Compost production trainings were also held in several communities within the municipality, with the support of project staff and the extension officers of MOFA; the main participants in these community trainings were men. Persons enrolled at the training centre undertook three modular curricula that entailed knowledge and skills learning in relation to organic waste collection, com-post production and entrepreneurship. The project registered a total of 65 persons who reported voluntarily to the training centre to be trained.

However, only 31 completed the training and were awarded certificates. The high dropout rate was a result of various issues. Some of the attendees thought the process was tedious, but many of them dropped out because they were not satisfied with the financial incentive given to them to support their travel to and from the training centre.

Fig. Trainees using tricycle for waste collection

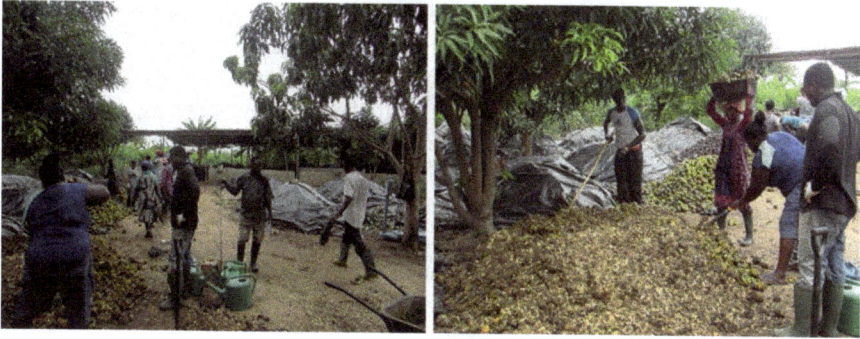

Fig.   Trainees undergoing compost production training

5.3   Engagement of Market Traders and Waste Segregation

With the support of the Environmental Health Officer of the municipality, the market queens took the lead in mobilising and identifying the best locations to site waste bins for temporary storage of their segregated food waste generated in the market. In one of the markets, the leaders took full responsibility for the protection and maintenance of the bins. They also made sure recalcitrant traders were not allowed to dump mixed waste into the bins. The commitment levels of the market women were impressive and within a few months of project implementation, many of the traders began to show consistency in segregating organic waste from their waste stream, though a few still remain reluctant to change their practice. The waste segregation is continuing in two of the markets, with the programme being disrupted in one of the markets due to market reconstruction and the other market dealing with leadership challenges.

5.4   Farmer Engagement and Training

Over 1750 farmers were provided with the technology for composting farm waste by the agricultural extension officers. Forty farmers are experimenting with farm waste composting on their farms. However, the purchase of compost by farmers is currently low. Farmer engagement and training activities were led by the agricultural extension officers. A household survey was conducted in October 2017: 175 farmers were randomly sampled from four working zones of the district office of MOFA using the purposive random sampling method, to assess farmers' perceptions of compost and its adoption level.

Almost one-third of the farmers surveyed (29%) were of the ages of 51–60 years, with only 2% of them above 70 years. The major crops cultivated by the farmers are cassava (*Manihot esculenta*), maize (*Zea mays*), garden eggs (*Solanum melongena*), pineapple (*Ananas comosus*), okra (*Abelmoschus esculentus*), tomatoes (*Solanum lycopersicum*) and pepper (*Capsicum*). The study revealed that over 80% of farmers were aware of the benefits of the use of compost. They stated benefits such as soil moisture retention, improvement of soil nutrients and increase in crops' shelf lives, amongst others. However, out of a total of 175 farmers interviewed, only 23.4% of them have adopted and are using compost, with the majority not using it. Questions on why they were not using compost were posed to people who had not adopted compost. Analysis of the factors influencing the low compost adoption and utilisation rate by farmers revealed factors such as the labour-intensive nature of compost preparation, a general lack of interest, the use of fertiliser and a lack of water for making compost, raw materials and knowledge on compost preparation. Also, access to extension services and land size were found to significantly influence compost adoption.

5.5   On-Farm Experimentation

Two field experiments were conducted at the University of Ghana research farm to evaluate the influence of compost and nitrogen fertiliser on plant height yield and biomass of maize (*Zea mays* L.). The study area was within the coastal savannah ecological zone of Ghana. The experiments were run during the minor season in 2016 and major season in 2017. The test crop was maize. Four concentra-tions of nitrogen (N) were considered with three levels of compost during the major season. However in the minor season, two levels of compost were applied (0 and 4000 kg/ha). Basal application of phosphorus (P) and potassium (K) was carried out for all plots; N was split-applied to respective plots and compost was applied to respective plots. Initial soil sampling was carried out before the establishment of trail.

The experimental design was a randomised complete block design, with 12 and 8 treatments and 3 replications during the major and minor seasons, respectively. Each treatment was separated from the other with a 1 m buffer strip (alley) between plots and a 1.5 m buffer strip between replication, to prevent cross-contamination between treatments.

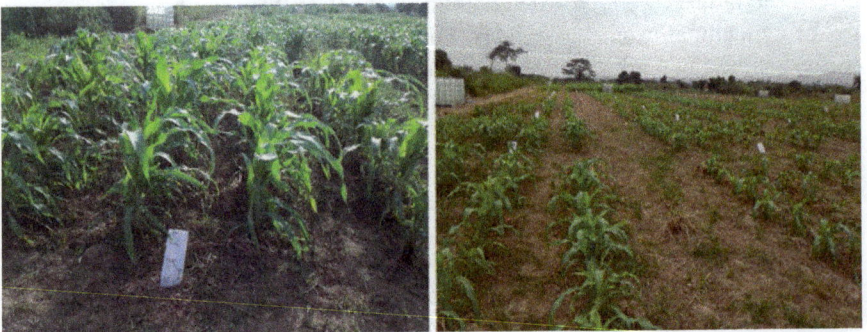

Fig.   Farm experimentation

Treatment combination	N (kg/ha)	C (kg/ha)
N1C1 (control)	0	0
N2C1	40	0
N3C1	80	0
N4C1	120	0
N1C2	0	4000
N2C2	40	4000
N3C2	80	4000
N4C2	120	4000
N1C2	0	6000
N2C2	40	6000
N3C2	80	6000
N4C2	120	6000

Table Treatment combination used in field experiment in 2017 major season

The results of the field experiments showed that the application of compost increased plants' growth, with a greater plant height in plots which received compost compared with those without compost application. The increase was greater during the major season compared to the minor season.

Generally speaking, nitrogen significantly ($P < 0.05$) increased the grain yield of maize up to 80 kg N/ha. Grain yield responded positively to N application, with yields ranging from 1124 kg/ha in N1C1 (control) to a maximum yield of 3651 kg/ha in N4C3 (120 kg N/ha and 6000 kg C/ha) during the major season. There was a significant increase ($P < 0.01$) in grain yield when N was applied, irrespective of the application of compost. Significant ($P < 0.05$) interactive effects of N and compost on grain yield were also observed. Similarly, nitrogen fertiliser and compost significantly increased the total biomass accumulation. The application of 6000 kg compost increased total biomass by 28% compared to treatment without compost application. In addition, compost significantly increased the total nitrogen uptake. The logistic (logit) model was employed to analyse the data due to the nature of the variable: whether farmers perceived compost to be good and had adapted or otherwise. For such a dichotomous outcome, the logit model is the most appropriate analysis tool. The logit model considers the relationship between a binary dependent variable (compost adoption) and a set of independent variables, whether binary or continuous.

5.6   The Business Model

Based on the assumption that the trainees had received comprehensive training in waste collection, composting and entrepreneurship, Business Model 2 was selected by the first batch of trainees for experimentation. They were responsible for

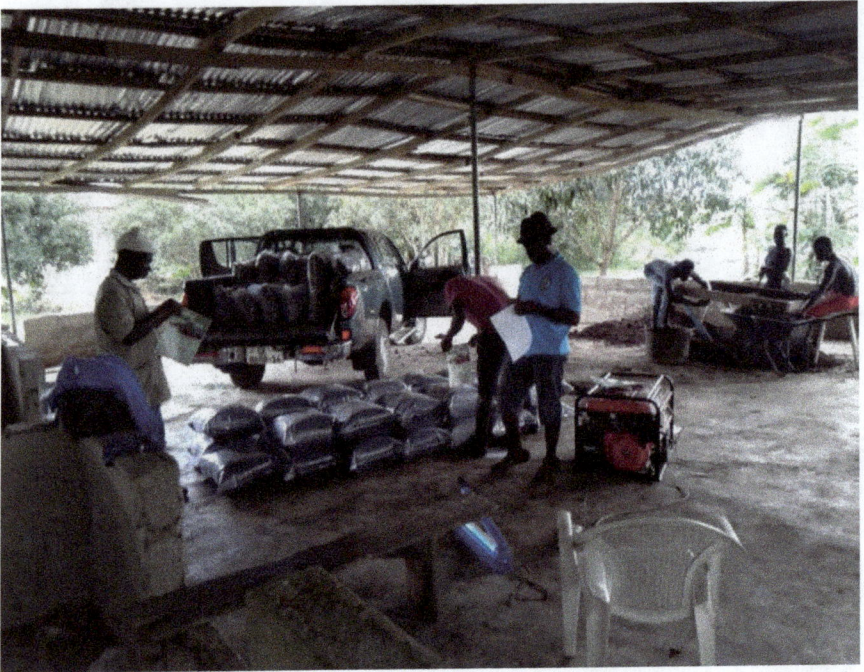

Fig. Youth group busy sieving and bagging compost for delivery

collecting, processing and producing compost, creating awareness of the product and marketing the product. After 3 months of experimentation it was difficult for them to sell their product because the farmers who were the initial target customers were not eager to use compost, i.e. there was no demand for the compost. Based on the experience of the first batch, the second batch of trainees experimented with Option 3.

A small-scale compost production business venture was set up by a six-member group of trainees, who have taken it up as an income-generating activity. Also, one of the trainees also set up a production unit in his home. The project provided support for the trainees by offering them tricycles on hire purchase and providing them with basic tools and technical support. To date, they have produced about 78 tonnes of compost and have sold compost worth about 2500 USD.

6 Key Outcomes and Achievements

Through the UOWIAP project, organic waste value chain opportunities have been created in the municipality, particularly amongst the public. Youth groups who engaged in the compost production and sale are enthused about the opportunities.

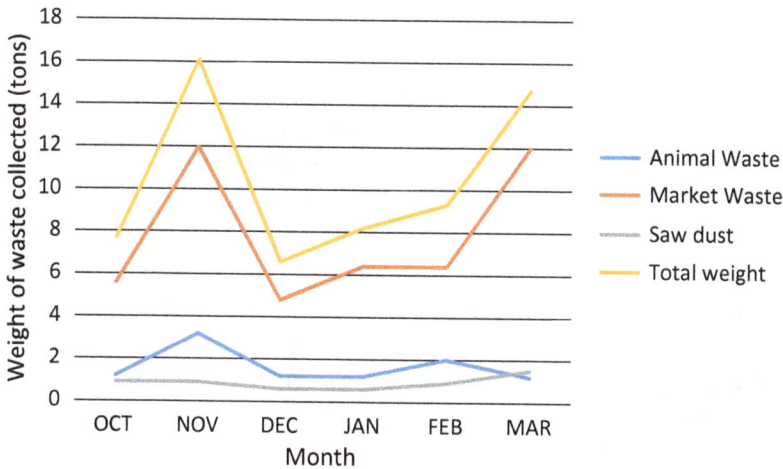

Fig. Weight of waste collected by youth group over a 6-month period between October 2017 and March 2018

Between October 2017 and March 2018, one of the groups had diverted about 62 tonnes of organic waste from landfill to compost production .

The general feedback from clients who purchased compost was very positive. The following are examples of comments received:

"Your product is much better than that from a common brand of compost on the market".
– comment from a grounds supervisor.
"Make sure you always maintain the quality of your compost, it is very good."
"I grow oriental vegetables in my home garden and had good yield when I used your compost."

The visibility of the project is also increasing as various stakeholders continue to pay visits to the project site to see the project. Visitors to the sites include a pawpaw exporters association, real estate agencies, other university units such as the West African Centre for Crop Improvement and a group of 12–15 participants of the Valorisation and Innovation in Africa (VIA Water) project from about six countries.

The fact that a few farmers have begun experimenting with compost production on their farms is a success story for the project. Farmers in the municipality have been trained on the benefits of compost use for soil health and nutrient. The project has also increased the number of farmers who have adopted compost use and production. Farm experimentation increased by between 12% and 20%.

Knowledge co-creation efforts were very helpful in strategy development, especially with the waste collection and awareness creation components of the project. The direction given by traders informed collection frequency and time and the establishment of a cordial working relationship between the youth group and the traders.

7   Challenges

Driving behavioural change towards good environmental sanitation is a complex issue that needs innovative strategies and intense stakeholder engagement to achieve success. Outlined below are some pertinent challenges faced in the implementation of the project.

7.1   Stakeholder Engagement with Government Agencies

The experience with engaging governmental agencies faced challenges, which varied from issues with project staff mobility to the business-as-usual approach to work. Within the project period, about four government agency representatives on the project were transferred to other jurisdictions and this hindered project activity continuity, especially in the area of monitoring and supervision and further engagement. Secondly, certain components of the study were to be incorporated into government staff's normal programmes of work, which also significantly delayed implementation. For instance, the Agricultural Extension Services for many months did not receive their budgetary allocations and hence on many occasions did not have funds to fuel their vehicles and motorbikes to engage in their routine jobs. The bureaucracy and the general business-as-usual approach to work within the governmental institutions also delayed processes.

7.2   The Low Level of Awareness of the Value of Compost

There were misconceptions amongst the farmers about compost value as regards soil quality. Many of the farmers considered the compost's effect on soil to be slow and did not support the fast yields expected, as with the application of chemical fertilisers. The value of compost as a soil conditioner was only marginally appreciated. The adoption of new technologies by farmers generally takes time. Likewise, the practicality of composting on farms needs further investigation. One of the major concerns raised by farmers in relation to composting is the availability of water. Farm composting can be practised during the non-farming season and this coincides with the dry season.

Because of the low level of awareness and use of compost nationwide, there is low demand for compost. Also, the limited market space for organically produced foods is a disincentive, and farmers are forced to sell at the same price as inorganic fertilised food. This situation needs to be addressed through massive campaigns (education and awareness creation) to create demand for organic foods.

7.3 Practice of Non-segregation of Waste from Source

Non-segregation is a major drawback to waste valorisation. In Ghana, the practice of waste segregation occurs on a very limited scale and the large volumes of municipal waste generated are mixed. This problem is related to the fact that municipalities have no systems in place that promote waste segregation from source and the behaviour of producing mixed waste is entrenched in many Ghanaians.

8 Sustainability Pathways

To ensure project sustainability, a strong commitment to promote waste segregation is required from the Municipal Assembly. The Municipal Assembly is responsible for waste management and has the mandate to create an enabling environment to promote the expected behaviour in order to promote sustainable waste management. They must invoke their powers to set the necessary by-laws for waste segregation from source, and they must enforce them. The Municipality Assembly engages the private sector to collect waste from the municipality and therefore can direct the private sector through contractual agreements to begin the collection and treatment of segregated waste, whilst providing the necessary support for them.

There is also a need to create demand for compost products, such as organic foods. The post-project implementation phase is ongoing, with efforts taking place to develop a national compost use campaign. Through a partnership with the University of Oregon advertising team and other stakeholders, a prototype campaign has been developed to target middle-income urban dwellers to demand organic foods. Farmer education and sensitisation to use compost is also ongoing through the MOFA extension services. The project is also working to engage further with MOFA on the government's "planting for food and jobs" initiative.

9 Conclusions

The educational and awareness programme undertaken by the project, against the backdrop of the current knowledge and experiences of the stakeholders with respect to the mounting municipal waste menace, has heightened the need to find appropriate solutions to the waste challenge. This, together with the right support for composting, would help in addressing the problem. The experiences of the project suggest the need to build the capacities of compost producers, farmers, market participants and other parties on sustainable municipal organic waste management. The following measures are necessary:

1. The Municipal Assembly, as a matter of urgency, must begin to drive waste segregation from source by putting in place necessary by-laws and providing resources for enforcement. They should also engage the private sector in collecting segregated waste.
2. There should be continuous sensitisation of traders at the local markets on the value of segregating the waste they generate. In this regard, the municipal authorities and market management should make available prescribed containers at designated sites at the market place and ensure that they are used appropriately.
3. There should be intensified and wider educational coverage of compost use amongst farmers in the urban and peri-urban areas.
4. The unemployed youth should be given continuous technical training in compost production and compost market opportunities should be created for them.
5. Extension services should be provided with the necessary logistics and resources.

Acknowledgements The authors are sincerely grateful to the Ministry of Foreign Affairs of the Netherlands Government, through the Netherlands Organisation for Scientific Research (NWO-WOTRO), for funding this study. We are also thankful to all stakeholders engaged, whose contributions have impacted on the success of this project. Our sincere gratitude goes to the University of Oregon advertising faculty and students, who have partnered with the project to develop a compost campaign. Also acknowledged are the support staff of the Institute for Environment and Sanitation Studies, particularly Mr. Logosu, Mr. Lar and Mr. Acheampong, and the laboratory technicians who worked tirelessly on this project.

References

Bonti-Ankomah, S., & Yiridoe, E. K. (2006). *Organic and conventional food: A literature review of the economics of consumers' perceptions and preference*. Final Report, Organic Agriculture Centre of Canada, Nova Scotia Agricultural College, Truro, Nova Scotia, Canada.
Cofie, O., Rao, K. C., Fernando, S., & Pau, J. (2009). *Composting experience in developing countries: Drivers and constraints for composting development in Ghana, India, Bangladesh and Sri Lanka*. Final Report for World Bank, International Water Management Institute (IWMI), Colombo, Sri Lanka.
Danso, G., Drechsel, P., Fialor, S., & Giordano, M. (2006). Estimating the demand for municipal waste compost via farmers' willingness-to-pay in Ghana. *Waste Management, 26*(12), 1400–1409.
Food and Agriculture Organization (FAO). (2004). *Rural households and resources: A pocket guide for extension workers, Socio-Economic and Gender Analysis Programme (SEAGA)*. Rome: Food and Agriculture Organization.
Gabbay, O. (2010). *Assessment of the Avenor pilot community composting facility and its adaptability in other Accra sub-metro areas*. Report prepared for CHF International Ghana.
Ga-West Municipal Assembly. (2010). *Ga-west medium term development plan 2010–2013*. Ga West Planning Department.
Ghana Statistical Service (2014). *2010 population and housing census*. Summary report of final results, Accra, Ghana.
Hoornweg, D., & Bhada-Tata, P. (2012). *What a waste: A global review of solid waste management* (Urban development series; knowledge papers no. 15). Washington, DC: World Bank.

Miezah, K., Obiri-Danso, K., Kádár, Z., Fei-Baffoe, B., & Mensah, M. Y. (2015). Municipal solid waste characterization and quantification as a measure towards effective waste management in Ghana. *Waste Management, 46*, 15–27.

Nartey, E. G. (2013). *Faecal sludge reuse in urban and peri-urban crop production.* A thesis submitted to the University of Ghana, Legon in partial fulfilment of the requirement for the award of Master of Philosophy degree in environmental science.

Starovoytova, D. (2018). Solid Waste Management (SWM) at a University Campus (Part 1/10): Comprehensive-review on legal framework and background to waste management, at a global context. *Journal of Environment and Earth Science, 8*(4), 68–116.

Vergara, S. E., & Tchobanoglous, G. (2012). Municipal solid waste and the environment: A global perspective. *Annual Review of Environment and Resources, 37*(1), 277–309.

Valuing Waste – A Multi-method Analysis of the use of Household Refuse from Cooking and Sanitation for Soil Fertility Management in Tanzanian Smallholdings

Ariane Krause

Abstract The starting point of this work is the intention of two farmers' initiatives to disseminate locally developed and adapted cooking and sanitation technologies to smallholder households in Karagwe District, in northwest Tanzania. These technologies include improved cooking stoves (ICSs), such as microgasifiers, and a system combining biogas digesters and burners for cooking, as well as urine-diverting dry toilets, and thermal sterilisation/pasteurisation for ecological sanitation (EcoSan). Switching to the new alternatives could lead to a higher availability of domestic residues for soil fertility management. These residues include biogas slurry from anaerobic digestion, powdery biochar from microgasifiers and sanitised human excreta from EcoSan facilities. Such recycling-driven approaches address an existing problem for many smallholders in sub-Saharan Africa, namely, the lack of soil amenders to sufficiently replenish soil nutrients and soil organic matter (SOM) in soils used for agricultural activity. This example from Tanzania systematically examines the nexus of 'energy-sanitation-agriculture' in smallholder farming systems. The short-term experiments demonstrated that all soil amenders that were analysed could significantly enhance crop productivity. CaSa-compost – the product of co-composting biochar with sanitised human excreta – quadrupled grain yields. The observed stimulation of crop yield and also plant nutrition is attributed to improved nutrient availability caused by a direct increase of soil pH and of plant-available phosphorus (P) in the soil. The assessment of the lasting soil implications revealed that CaSa-compost and biogas slurry both show the long-term potential to roughly double yields of maize. Corresponding nutrient requirements can be adequately compensated through residue capturing and subsistence production of soil

---

This article is an excerpt from the author's doctoral thesis (Krause 2018). The text is, therefore, partly revised and partly reprinted.

A. Krause (✉)
Leibniz Institute of Vegetable and Ornamental Crops (IGZ), Großbeeren, Germany
e-mail: Krause@igzev.de

amenders. The potential of CaSa-compost for sustainable soil fertility management is superior to that of standard compost, especially with respect to liming, replenishing soil P and restoring SOM. Biogas slurry, however, yields inferior results in all aspects when compared to compost amendments.

Keywords Biochar · Bioenergy · Biogas slurry · Compost · Ecological sanitation · Soil amendments · Soil improvement · Terra preta practice · Waste as resource

1   Introduction

1.1   *Sustainable Food Production and the 'Energy-Sanitation-Agriculture' Nexus*

Providing a growing population with healthy food from sustainable production systems is one of the most urgent contemporary challenges for global societies and science. In this context, the EAT-Lancet Commission on Food, Planet, Health concludes that a 'great food transformation' is urgently needed and demands 'an agricultural revolution', with a focus on diets and food production practices that will help to achieve the UN Sustainable Development Goals, as well as the Paris Agreement on Climate Change (Willett et al. 2019). Regarding anthropogenic activities impacting the 'planetary boundaries' (which define a safe operating space for humanity on Earth), Springmann et al. (2018) showed that only the combination of (1) ambitious technological improvements, combined with (2) dietary changes towards more plant-based diets, and (3) the reduction in food loss and waste has the potential to keep the environmental impacts of the food system at bay. Technological changes are in this case particularly relevant for the environmental domains of global nitrogen (N) and phosphorus (P) application (ibid.). Furthermore, the International Assessment of Agricultural Knowledge, Science and Technology for Development (IAASTD), also known as the World Agriculture Report, sees 'agriculture at a crossroads' and calls for focusing on efficient, small-scale agroecosystems with nutrient cycles that are as closed as practicably possible (McIntyre et al. 2009). Circular economies (CE) and material cycling within the agroecosystem are, therefore, necessary to stay within a safe operating space for food systems and represent agreed prerequisites for long-term global food production, which has been promoted inter alia by Willett et al. (2019), Springmann et al. (2018), the UN Food and Agriculture Organization (FAO) (2014), Lal (2006, 2009), La Via Campesina (2015), McIntyre et al. (2009) and De Schutter (2011).

Whilst smallholder farmers cultivate at least half of the world's food crops (Graeub et al. 2016), many smallholders in sub-Saharan Africa nevertheless lack the resources to sufficiently replenish soil nutrients and soil organic matter (SOM) in soils depleted by agricultural activity (Buresh et al. 1997; Markwei et al. 2008). This is why smallholders are often trapped in a vicious circle: soil acidity and P scarcity lead to insufficient production of food crops, which in turn leads to insufficient availability of residual matter for soil fertility management. Here, (the CE) approach and considering waste as a resource can be essential for closing material loops. As an example, waste streams from cooking and from sanitation can be employed in recycling-driven soil fertility management. Residues from cooking that can be recovered as resources for agriculture include ashes from the most common three-stone fires, biochar (i.e. char particles) from a kind of improved cook stove called a 'microgasifier' and biogas slurry from biogas systems. Biochar in particular is rich in carbon (C), and its recovery can therefore contribute to restoring SOM, whilst biogas slurry is valued as nutrient-rich fertiliser. In addition, bioenergy can also be applied to sanitation processes in order to destroy or deactivate pathogens from human excreta. When managing human excreta, preventing the transmission of disease is an essential element of EcoSan. Sanitation, therefore, needs to take place at as early a stage as possible during the process (World Health Organization (WHO) 2006). Technological sanitation alternatives that make it possible to consider human excreta as a resource, rather than as waste, include the urine-diverting dry toilet (UDDT), which collects human excreta, as well as thermal sanitation via pasteurisation and composting to properly treat the collected matter. Once sanitation has been completed, urine and faeces constitute a valuable resource for recycling plant nutrients, including N, P, potassium (K) and micronutrients (Esrey et al. 2001). Such recycling practices around the 'energy-sanitation-agriculture nexus' thus create a link between the use of sustainable technologies for energy and sanitation services and the recovery of waste resources from cooking and EcoSan to produce recycling fertilisers for agriculture. However, within the broad field of recycling fertilisers, there is still a lack of knowledge on applicability and the agronomic and ecological potential in agricultural and horticultural practices, which is particularly required when promoting CE to vulnerable smallholder communities.

1.2   Objectives and Outline

Against this backdrop, it was the objective of the present study to investigate an applied example of integrated resource management around the energy-sanitation-agriculture nexus, which is realised in Karagwe, Tanzania, and which combines the implementation of cooking and sanitation facilities and the recovery of residues. I was particularly interested to study the potential of recycling-driven soil fertility management to fill existing fertiliser gaps in smallholder farming.

In this article, I firstly describe the study area and thereafter introduce the local projects and their underlying technologies. I then explain the methodologies I applied to analyse the agronomic potential of linking waste streams from cooking and sanitation with agriculture. In Sect., I summarise and discuss results from an exploratory study conducted on-site and from a model-based simulation of long-term effects. In Sect. , I present and discuss opportunities and challenges for the real-world application of the studied recycling practices, before I close with presenting the main conclusions of my work (Sect.).

2 Description of the Study Area

2.1 Location and Climate

Karagwe is one of eight districts in Kagera Region in northwest Tanzania (01°33′ S, 31°07′ E, alt. 1500–1600 m.a.s.l.). Kagera is part of the Lake Victoria Basin and is located near to the volcanic areas of the East African Rift Zone. The typical terrain of Karagwe District consists of hills and valleys. Rainfall is bimodal (March–May and October–November) and varies between 500 and 2000 mm/year, whilst mean temperatures range from 20 to 28 °C during the day. This semi-arid and tropical savanna climate (according to the Köppen-Geiger climate classification in Peel et al. 2007) allows twice-yearly harvests for most annual crops. The regional economy is dominated by smallholder agriculture, with about 90% of households selling agricultural products grown on their farms (United Republic of Tanzania (URoT) 2012).

2.2 Rural Livelihoods in Karagwe District

On average, a Karagwe household comprises six individuals (URoT 2012). The main resource for cooking is biomass: 96% of households use firewood, whilst about 3% use charcoal (ibid.). Sanitation facilities are mainly pit latrines: 88% of households use standard pit latrines, compared to just 4% of households using improved pit latrines; only 1% of households possess a system of flush or pour toilets in combination with septic tanks, and 6% do not have any toilet facilities at all (ibid.). These patterns, alongside population growth, have increased pressure on natural resources, including soil, open and running water, forests, etc., over the last three decades (Ogola 2013; Rugalema et al. 1994).

2.3 The Agroecosystem in Karagwe

Most farms in this region have less than one hectare of planted land and are fully or partly subsistent (URoT 2012). A farm usually comprises a large, permanently cultivated plot (a shamba, also known as a kibania) and small fields planted with annual crops (msiri, also known as omusiri or kikamba) (Baijukya and de Steenhuijsen

Piters 1998; Rugalema et al. 1994). The shamba (meaning 'field' in Swahili; it is also translated as a 'banana-based home garden') typically features a multi-layer design with diverse crops: high-growing, shady perennial crops, such as banana plants or fruit and coffee trees, that are intercropped with low-growing, annual crops as cover crops, such as beans, cassava, wild varieties of African eggplant, etc. On msiri plots, annual crops such as maize, potatoes, cassava, vegetables, etc. are cultivated in proximity to each other ('intercropping') (ibid.). The shamba is the most important type of land use for agriculture in Kagera, accounting for >40% of total agricultural land (URoT 2012). Msiri cultivation, meanwhile, is implemented on about one-fifth of the farmland (ibid.). Regarding soil fertility management, the shamba usually receives much more attention compared to the msiri fields (Rugalema et al. 1994).

2.4 Soil Pre-conditions in Karagwe

Soils in Karagwe and Kagera are commonly predominantly Ferralsols and Acrisols (Kenneth et al. 2003; DePauw 1984) but can also be classified as Andosols (Batjes 2011; Krause et al. 2016). Like many smallholders in sub-Saharan Africa, Karagwe farmers are challenged by soil constraints, including (1) soil erosion, (2) soil acidity and (3) nutrient deficiencies – in particular the scarcity of plant-available P. In the past, nutrient input and output flows were balanced by effective material cycling (Rugalema et al. 1994). However, the growing market economy over the last decades has increased the export of nutrients from the region, especially in cash crops, such as coffee and banana (ibid.). As a result, the current nutrient uptake by crops often exceeds the input from fertilisers (Baijukya et al. 2006; United Nations Environment Programme (UNEP) 2007). With regard to the regional economy, the local soil is amongst the most important production factors for local communities, who consequently assign significant importance to the conservation and amelioration of agricultural soils. In addition, increasing fragmentation of farmland due to population growth places ever higher demands on the land for sustaining food security and farm incomes (Rugalema et al. 1994). As a consequence, there is a strong demand for an increase in agricultural productivity.

2.5 Soil Management Applied in Karagwe

According to Tittonell (2016), farmers in East Africa often pay special attention to a certain portion of their land where specific, highly valued crops are cultivated. This may result in what Tittonell has described as 'islands of soil fertility'. Many farmers in Kagera are proud of their typically well-managed banana-based home gardens (Rugalema et al. 1994). The shamba commonly receives sufficient inputs of agricultural and domestic 'waste' as it directly surrounds the farm. The msiri, situated on close fields, often receives fewer inputs and is, therefore, typically more adversely affected by nutrient depletion. In regard to balancing inputs and outputs of soil nutrients for the shamba, Baijukya and de Steenhuijsen Piters (1998) showed that structurally poor households without cattle are most affected by declining soil fertility. As countermeasures, Baijukya and de Steenhuijsen Piters (1998) made recommendations including the increased use of compost and effective recycling of all household refuse, including human excreta. Standard compost in Karagwe contains a mixture of fresh and dried grasses, ash, kitchen waste and, depending on the availability, animal manure or leftovers from brewery processes (Krause et al. 2015). In addition, water is added (if available) to improve moisture content. Composting is carried out in batches and usually takes around 3–6 months. Compost heaps are often placed in shallow pits under the shade of trees and covered with soil and grasses to mitigate evaporation. In addition to compost, the most common soil fertility management measures are carpeting with grasses and mulching with harvest residues.

3 Material and Methods

As indicated in the previous sections, there is, for multiple reasons, a strong demand for soil fertility management and appropriate soil amenders in Karagwe. In particular, msiri cultivation has not yet been fully researched. For this reason, I focus on researching realistic options for managing the local Andosol through the use of locally available residues from cooking and sanitation technologies implemented in the case study projects , which I introduce in the following section.

3.1 Case Study Projects and Technologies

Two farmers' initiatives in Karagwe and their German partners have recently developed a set of projects, which have two aims: (1) providing the local community with resource-efficient cooking and sanitation technologies, whilst (2) improving local crop production by rigorous on-farm recycling of residues from cooking and sanitation and recycling-oriented soil fertility management. The farmers' initiatives facilitate projects at a grass-roots level, namely, MAVUNO (Project for Improvement for Community Relief and Services) and CHEMA (Programme for Community Habitat Environmental Management). The three local projects are (1) Biogas Support for Tanzania (BiogaST), (2) Efficient Cooking in T anzania (EfCoiTa) and (3) Carbonisation and Sanitation (CaSa). The perspective adopted in the implementation has mainly focused on the household level, but it also looked at the institutional level. The projects' technologies are explained schematically and implemented as illustrated in Figs..

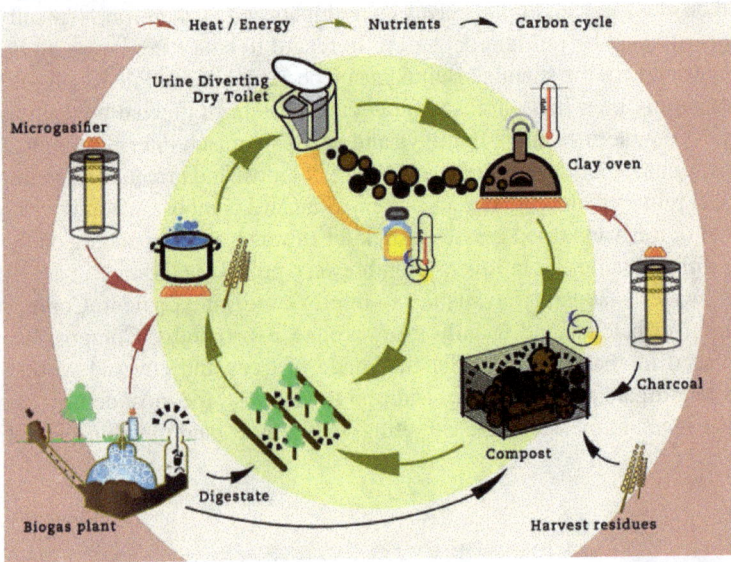

Fig. Intersectional resource management using residues from cooking and EcoSan for soil improvement in the context of smallholder farming in Karagwe.

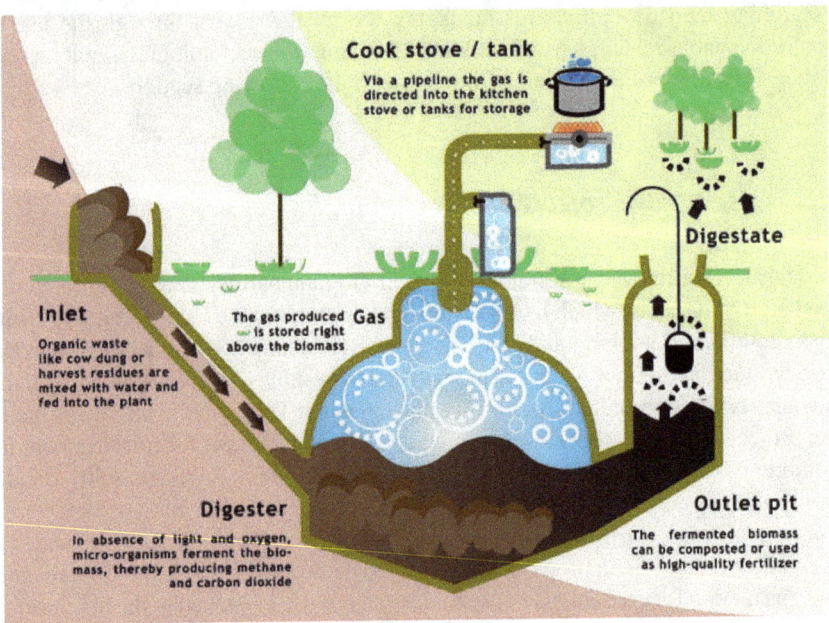

Fig. Working principle of a small-scale biogas digester, including subsequent use of the bio-gas for cooking and of the biogas slurry/digestate for fertilisation.

Fig. The biogas system analysed in the present work comprises a small-scale biogas digester (left and centre) and a LOTUS biogas burner (right). (Source: Schrecker 2014)

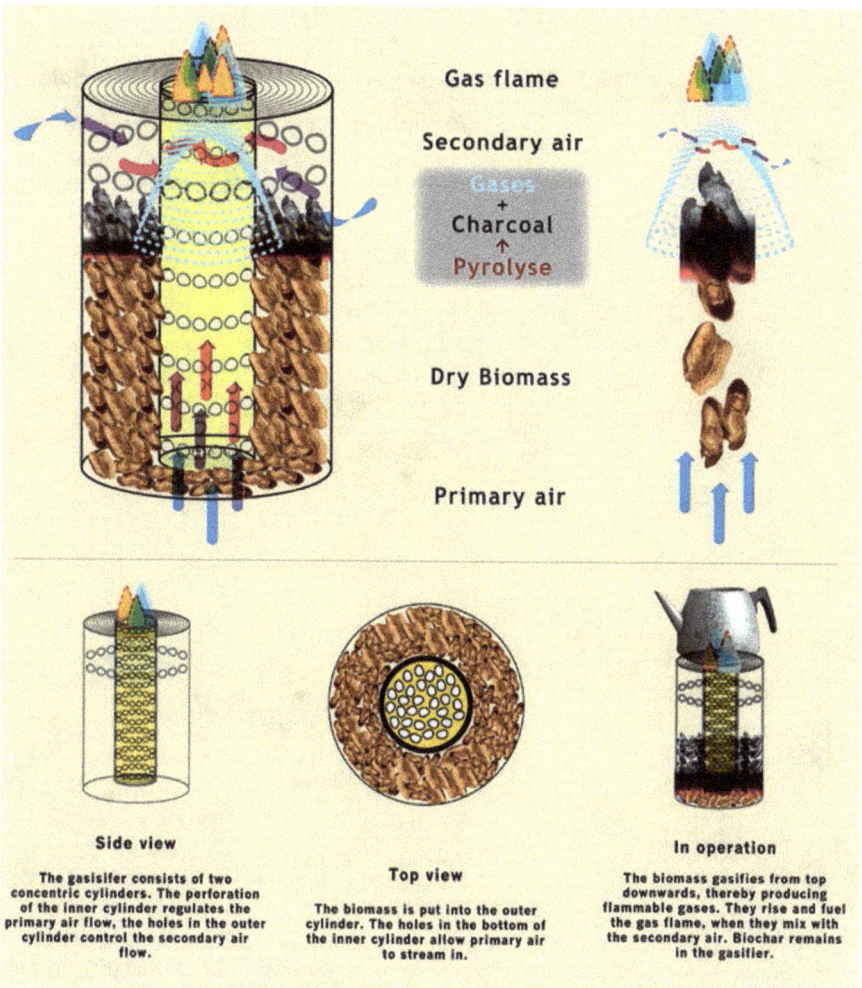

Gas flame

Secondary air

Gases
+
Charcoal
↑
Pyrolyse

Dry Biomass

Primary air

Side view

The gasisifer consists of two concentric cylinders. The perforation of the inner cylinder regulates the primary air flow, the holes in the outer cylinder control the secondary air flow.

Top view

The biomass is put into the outer cylinder. The holes in the bottom of the inner cylinder allow primary air to stream in.

In operation

The biomass gasifies from top downwards, thereby producing flammable gases. They rise and fuel the gas flame, when they mix with the secondary air. Biochar remains in the gasifier.

Fig. Working principle of a microgasifier used for cooking, including co-production of biochar. The example given is of a stove operating with biomass in small pieces, such as groundnut shells. (Infographic (CC) by Lusi Ajonjoli)

Fig. The microgasifier stoves analysed in the present work comprise an improved sawdust stove (left) and a TLUD stove (right). (Sketches and photographs by D. Fröhlich)

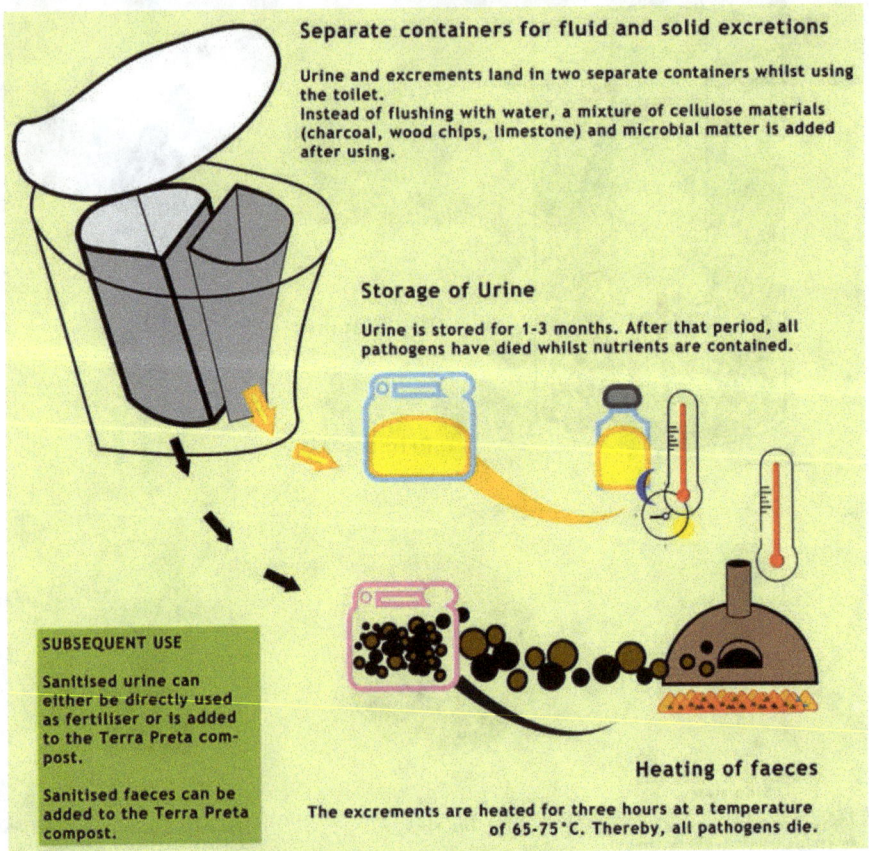

Separate containers for fluid and solid excretions

Urine and excrements land in two separate containers whilst using the toilet.
Instead of flushing with water, a mixture of cellulose materials (charcoal, wood chips, limestone) and microbial matter is added after using.

Storage of Urine

Urine is stored for 1-3 months. After that period, all pathogens have died whilst nutrients are contained.

SUBSEQUENT USE

Sanitised urine can either be directly used as fertiliser or is added to the Terra Preta compost.

Sanitised faeces can be added to the Terra Preta compost.

Heating of faeces

The excrements are heated for three hours at a temperature of 65-75°C. Thereby, all pathogens die.

Fig. Working principle of a UDDT, including possible paths for the subsequent treatment of urine and faeces, comparable to those analysed in the present study. (Infographic (CC) by Lusi Ajonjoli)

Fig. The CaSa approach to EcoSan analysed in the present work comprises a UDDT (left) and sanitation oven (right). (Photographs by A. Krause, sketches from CaSa (2012))

The BiogaST project aims to provide cooking energy from a biogas system comprising a small-scale biogas digester with a plug-flow fermenter using harvest and kitchen residues and a 'LOTUS' biogas burner.

The EfCoiTA project works with ICSs, such as top-lit updraft (TLUD) microgasifier stoves that operate with pieces of firewood or maize cobs and an improved sawdust microgasifier stove that utilises sawdust or coffee shells. Both the biogas system and the ICSs have been designed with the aim of reducing, or even substituting for, the use of firewood.

The CaSa project deals with EcoSan, including a squat-type UDDT and a clay oven for thermal sanitation of excreta via pasteurisation, heated with a microgasifier .

Overall, and if they are successfully implemented, it is expected that the introduced technologies will increase locally available soil amendments by applying (1) biogas slurry from anaerobic digestion, (2) powdery biochar from microgasifiers and (3) sanitised human excreta.

3.2   Methods Applied to Study Recycling-Based Soil Management Strategies

As part of my PhD research, I firstly characterised and assessed substrates derived from the case studies' pilot projects for agricultural use through laboratory analysis (Krause et al. 2015). These substrates included biogas slurry and CaSa-compost, the product of co-composting biochar with sanitised human excreta and other organic and mineral residues (e.g. harvest residues, grasses, brick particles, etc.; cf. Krause et al. 2015). Then I examined the short-term effects of the potential soil amenders on soil fertility and crop productivity in a practice-oriented field experiment carried out in Karagwe in 2014 (Krause et al. 2016). In this exploratory study, I used locally available substrates as soil amenders for the cultivation of locally grown and nutritionally desirable crop species, which I intercropped like common in msiri systems. I discussed differences observed in plant growth and crop nutrition, in relation to nutrient and water availability in the soil.

In addition, I also estimated and evaluated long-term effects of different fertilisation strategies on soil quality and the production of crops commonly cultivated in msiri systems. This assessment of lasting soil implications is based on using the 'Soil and Water Integrated Model' (SWIM) of the Potsdam Institute for Climate Impact Research (PIK – *Potsdam-Institut für Klimafolgenforschung*). The SWIM is a process-based, ecohydrological model, which integrates the impacts of climate and land management, or the 'way of farming', with hydrological processes, soil erosion, nutrient dynamics and vegetation (Krysanova et al. 2000). The model has already been calibrated for Tanzania, including the Kagera Region (Gornott et al. 2015, 2016). The timeframe of the simulation is usually two decades (the period from 1993 to 2012). The SWIM evaluation considers several output parameters, including (1) annual crop yields of maize grains removed from the field during two seasons [t/ha], (2) soil P [kg/ha] comprising labile P (P_{lab}) as mineral P in the soil solution readily available for plant uptake and (3) soil N [kg/ha] comprising organic N (N_{org}) that is readily mineralisable and NO_3^- as mineral N (N_{min}) in the soil solution that is readily available for plant uptake. Input data to SWIM was based on two other studies that I conducted earlier by applying the material flow analysis method (cf. Krause and Rotter 2017) and material flow analysis in combination with the soil nutrient balances (cf. Krause and Rotter 2018). In total, the SWIM evaluation com-pared four scenarios, reflecting (1) the 'current state' of soil fertility management with limited fertiliser application (AM1) and (2) 'improved' soil fertility management with fertiliser applications based on using residues from cooking and sanitation (AM2–AM4). In addition, and in accordance with local practices, carpeting with grasses and mulching with harvest residues was applied as a 'basic practice' in all scenarios.

Table. Scenarios analysed in the evaluation of long-term effects on soil fertility: scenarios AM1–AM4 reflect d ifferences i n s oil m anagement p ractices t hrough t he u se o f t he a nalysed substrates as fertilisers in the msiri

Abbr.	Name	Fertiliser application for maize
AM1	'Current state'	None
AM2	BiogaST scenario	Biogas slurry and urine
AM3	'Optimistic' CaSa scenario[a]	CaSa-compost and urine
AM4	'Pessimistic' CaSa scenario[b]	CaSa-compost and urine

The abbreviation 'AM' reflects the 'agroecosystem' of a msiri
[a]With comparatively higher yield assumptions based on empirical results gained with CaSa-compost in Krause et al. (2015)
[b]With comparatively lower yield assumptions based on empirical results gained with the standard compost in Krause et al. (2015)

4   Discussion of Results

This chapter summarises and discusses the results of the three evaluation methods applied. Key findings focusing on the potential for soil fertility management in Karagwe are summed up in 4.4.

4.1   Laboratory-Based Characterisation of Locally Available Substrates

With respect to the assessment of locally available biogas slurry, standard compost and CaSa-compost for their fertilisation potential, the analysis of nutrient concentrations and nutrient availability revealed the following initial findings (cf. Krause et al. 2015): all treatments tested are characterised by sufficient nutrient concentrations for appropriate plant fertilisation and adequate nutrient ratios to avoid immobilisation of nutrients in the soil. They show good liming potential compared to other soil amendments, such as poultry or cattle manure, ammonium sulphate, urea, etc. The CaSa-compost shows an outstanding fertilisation potential, with the highest concentration of all analysed nutrients: for example, P concentration in CaSa-compost is 1.7 g/dm^3 compared to 0.5 and 0.3 g/dm^3 in standard compost and biogas slurry, respectively. Also, regarding its liming effect, CaSa-compost is outstanding, with 7.8 kg CaO/kg N, compared to 2.6 and 3.4 kg CaO/kg N of the standard compost and biogas slurry, respectively.

4.2   Empirical Study of the Use of Locally Available Substrates as Soil Fertility Improvers

After the initial characterisation of substrates, a subsequent one-season experiment provided empiric evidence of the immediate effects on soil fertility and biomass growth (Krause et al. 2016). With respect to soil physicochemical properties, I found that the availability of nutrients in the soil improved, in particular, after amending the Andosol with CaSa-compost, due to (1) significantly higher addition of total P (P_{tot}) with comparable doses of N_{min} and (2) a significant effect on soil pH, which is attributed to biochar contained in CaSa-compost. Levels of soil P rose significantly after the experiment. In soil amended with CaSa-compost, soil P reached a concentration of calcium acetate lactate soluble P (P_{CAL}) of 4.4 mg/kg, compared to 0.5 mg/kg in unamended control plots. No significant effect was observed on the availability of soil water, on concentrations of total organic carbon or on effective cation exchange capacity for any of the amendments tested. With respect to biomass growth and crop productivity, I found that biomass of maize increased in particular, when using CaSa-compost. Grain yields of maize in soil

Fig. Progress of the field experiment – 60 days after initiating the experiment with sowing of maize: an untreated plot ('without'), compared to plots amended with biogas slurry, standard compost and CaSa-compost. (Source: photographs taken by A. Krause on 2 June 2014)

treated with CaSa-compost, standard compost or biogas slurry increased to about 400%, 290% and 240%, respectively, when compared to yields in unamended soil. Biomass and yields of beans and onions significantly increased in the case of soil amended with CaSa-compost and standard compost only, whilst plant growth of cabbage clearly increased for all three amendments tested. Finally, and with respect to plants' nutrient status, I found that the total uptake of nutrients (N, P, K, calcium (Ca), magnesium (Mg) and zinc (Zn)) into maize grains increased significantly for all treatments in particular, for plants grown in plots amended with CaSa-compost. Primary response of maize plants is related to mitigated deficiency of soil P.

4.3 SWIM

The SWIM-based evaluation of the long-term effects of the recycling practices on crop yields and soil nutrients revealed the following findings. With respect to crop growth, SWIM predicts that annual maize yields would slightly, but continuously, decrease over a period of two decades under current soil management, as the linear

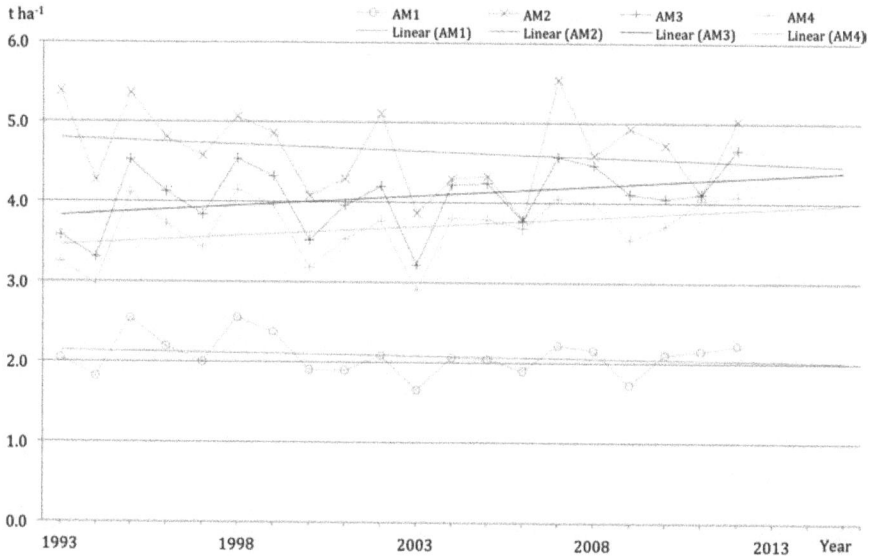

Fig. Temporal shifting of annual crop yields from the SWIM analysis: calculated annual crop yields from the SWIM for the period 1993 until 2012

trend line in Fig. indicates for AM1. Likewise, yields calculated for the BiogaST scenario decrease over the timeframe simulated, whilst grain yields estimated for the case of CaSa scenarios clearly increase over two decades. Furthermore, according to SWIM, yields of maize grains more or less double in scenarios AM2–AM4, compared with the baseline scenario AM1. However, (long-term) yields estimated with SWIM for BiogaST and CaSa scenarios increase to a lesser extent when compared with (short-term) empirical findings. Mean values of scenarios AM2, AM3 and AM4 are about 85%, 45% and 60% of annual yields projected from empirical data, respectively. Yields simulated for AM1 are, meanwhile, comparable to those realised in unamended soil during the field experiment (Krause et al. 2016) and fit with average grain yields reported for the region (Statistical Database of the FAO (FAOSTAT) 2012; Kimaro et al. 2009; Mourice et al. 2014). Yields simulated for scenarios AM2–AM4 are comparable to approximate yields required to reach world food security by 2025–2050 (Lal 2009).

Overall, according to SWIM, yields are higher for the case of using biogas slurry compared to using CaSa-compost, which is inconsistent with my empirical findings. I assume that SWIM overestimates higher inputs of Nmin with biogas slurry compared to CaSa-compost. By contrast, the residual effects of Norg applied and higher inputs of Ptot with CaSa-compost compared to biogas slurry are probably underestimated. Nonetheless, I observed empirically that maize plants directly react, in particular, to improved P fertilisation (Krause et al. 2016).

With respect to soil nutrient statuses, time series for Plab in the soil (Fig.) show a slight decrease of contents in AM1, which means that P is continuously depleted under current state conditions. However, in the BiogaST scenario, the

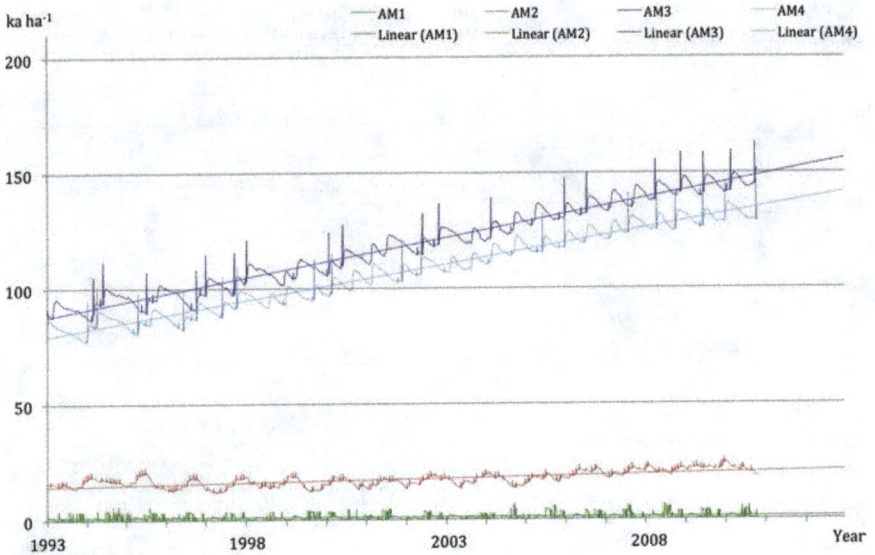

Fig. Temporal shifting of content of P_{lab} in the soil from the SWIM analysis: calculated con-tent of P_{lab} in the soil from the SWIM analysis for the period 1993 until 2012

content of Plab in the soil slightly increases, whilst the increase is clearly superior in both CaSa scenarios.

SWIM further projects a clear increase of N_{org} in the soil for CaSa scenarios, whilst in the BiogaST scenario, N_{org} only slightly increases over two decades. Under current soil management practices, however, the level of Norg in the soil remains constant over time. Simulated concentrations of NO_3^- in the soil fluctuate widely over the two decades, as was to be expected, pursuant to Finck (2007). The mean over two decades, simulating BiogaST and CaSa scenarios, is nearly seven and four times higher compared to the current state, respectively. Hence, according to SWIM, amending the soil with CaSa-compost offers outstand-ing potential in regard to continuously replenishing the soil nutrient statuses of N_{org} and P_{lab}.

With annual depletion or replenishment rates, on the one hand, and certain target values of soil P, on the other, it is possible to estimate potential timeframes for P replenishment. I found that a period of 30–300 years of continuously amending the local soil with CaSa-compost is required to reach ordinary benchmarks of extract-able soil P ranging from 100 to 800 kg/ha (Finck 2007; Landon 1991). Meanwhile, more than 700 years of constant CaSa practice are required to reach an extremely high, Terra Preta-like concentration of about 2500 kg/ha (Falcão et al. 2009). However, when using biogas slurry, the timeframe expands from several centuries to several millennia, depending on the target value. (Contents refer to a 1 m layer of topsoil.)

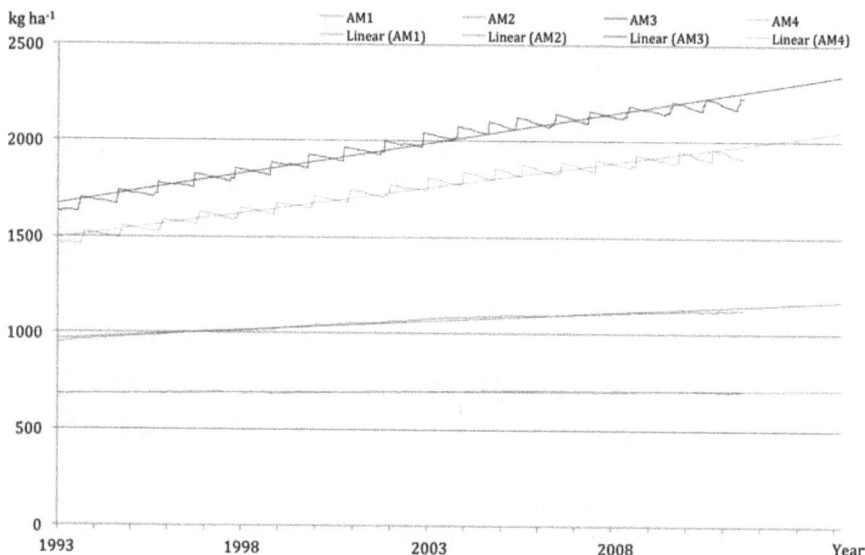

Fig. Temporal shifting of the content of N_{org} in the soil from the SWIM analysis: calculated content of N_{org} in the soil from the SWIM analysis for the period 1993 until 2012

4.4 Soil Fertility Management Around the Energy-Sanitation-Agriculture Nexus

The evaluation of opportunities and challenges when using substrates analysed for 'sustainable' soil fertility management in Karagwe focuses on the potential for (A) replenishing soil P, (B) mitigating soil acidity, (C) restoring SOM, (D) sequestering C and (E) increasing crop yields. In the following paragraphs, I briefly summarise the most relevant findings of my research with regard to this potential.

(A). A direct increase of soil P is practically possible with the addition of CaSa-compost at a rate of about 140 kg P_{tot}/ha (Krause et al. 2016). Adding biogas slurry or Karagwe standard compost at lower supply rates of 40 or 70 kg P_{tot}/ha, respectively, is insufficient to increase concentrations of soil P over the course of one season (ibid.). Nevertheless, on a farm level, the theoretic potential for annual P replenishment rates ranges between 20 and 60 kg P_{tot}/ha (Krause and Rotter 2018). The P application rates estimated for CaSa-compost and standard compost are similar and are in the upper half of the range presented. The potential of regular annual P applications is, therefore, sufficient for P fertilisation and P replenishment, pursuant to the recommendations of Buresh et al. (1997) and Nziguheba (2001). Potential P applications with biogas slurry, however, barely meet the minimum demand for P fertilisation on degraded soils with strong P fixation characteristics (ibid.). According to SWIM, long-term amendments of CaSa-compost demonstrate the clear potential to steadily increase soil P and therefore remedy P scarcity. This is not, however, the case when using biogas slurry. In practice, adequate levels of soil P support adaptation to, and mitigation of, the effects of climate change as a sufficient supply of soil P helps plants to root more deeply. This, in turn, makes crops less vulnerable to drought (Batjes and Sombroek 1997).

(B). A direct increase of soil pH, i.e. after a one-off soil amendment, is only possible with CaSa-compost with a liming potential corresponding to about 2000 kg CaO/ha (Krause et al. 2016). Amending the soil with biogas slurry or Karagwe standard compost, in application rates equivalent to liming with 300 or 700 kg CaO/ha, respectively, is not sufficient to increase soil pH over the course of a single season (ibid). The theoretical annual liming potential estimated in Krause and Rotter (2018) indicates that both CaSa-compost and Karagwe standard compost are feasible for maintaining soil pH (Finck 2007), whilst biogas slurry only fulfils the minimum requirements for liming, pursuant to Horn et al. (2010). Overall, acidity management through liming is an important soil management practice in order to strengthen nutrient-cycling processes (Batjes and Sombroek 1997). This means that an increase in soil pH through liming, as demonstrated especially for CaSa-compost, promotes nutrient availability in the soil and, thus, plant uptake of P and N. As a consequence, on-farm nutrient recycling becomes more efficient and agricultural activity more productive.

(C). In my experiment, none of the tested soil amendments practically altered total C content of soil over the course of a single cropping season after the application of 150 or 500 g C/m^2 with either biogas slurry or composts (Krause et al. 2016). The theoretical annual potential C contents in CaSa-compost and biogas slurry, meanwhile, are sufficient for restoring SOM consumed during the cultivation of maize in a field with areas of about 0.2 and 0.1 ha, respectively (Krause and Rotter 2017). Using potentially available CaSa-compost or standard compost for soil fertility management demonstrates that it is theoretically possible to restore sufficient C to the soil to surpass the humus consumption of C of those crops grown in msiri fields (Krause and Rotter 2018). In contrast, the C contained in biogas slurry barely balances SOM consumed during crop cultivation. To sum up, these findings indicate that only compost amendments demonstrate the theoretical potential to replenish SOM and also to sequester C in the long term; biogas slurry does not.

(D). Amongst the substrates analysed, CaSa-compost is particularly promising for sequestering C, due to the content of biochar recovered from household cooking and sanitation. Biochar has the potential for C sequestration due to the following factors: (1) it originates in renewable biomasses (Christensen et al. 2009); and (2) it is characterised by relatively recalcitrant organic compounds, which promise the long-term stability of biochar in the soil (Lehmann and Joseph 2009). The context of the present analysis further promotes C sequestration, as the local soil is known for its outstanding capacity to accumulate organic C. Andosols tend to protect organic matter from degradation by forming either metal-humus (i.e. often Al-Fe) or allophane-organo complexes (Zakharova et al. 2015). Therefore, and according to Chesworth (2008),

Andosols have the potential to act as CO_2 sinks. To the best of my knowledge, however, long-term studies observing the effect of biochar amendments on SOM in tropical Andosols do not exist. I am therefore unable to quantify the general potential for C sequestration with existing data, and any further discussion would enter into the realms of speculation. Nonetheless, I may at least compare my modelling results with the data available from short-term empirical studies in the region on biochar application and its effects. In the field trial, I observed that adding biochar to the local soil at a rate of around 2 kg/m² had no significant effect on soil total organic carbon (TOC) content (Krause et al. 2016). The theoretical potentials for biochar amendments in the msiri are estimated for annual or triennial biochar application rates of 0.3 or 0.8 kg/m², respectively, which correspond to C additions of approximately 0.2 or 0.6 kg C/m², respectively (Krause and Rotter 2018). These biochar amendments are, however, significantly lower than those recommended by Liu et al. (2012), or those in the practical experiences of Kimetu et al. (2008).

(E). Finally, all tested soil amendments can directly alter biomass production and crop yields (Krause et al. 2016). In the case of maize, CaSa-compost has the potential to quadruple grain yields in the short term (ibid.; Sect.). In the long term, and according to SWIM (Sect.), CaSa-compost and biogas slurry both have the potential to roughly double yields of maize grains. The empirical and analytical findings regarding the potential effects of CaSa-compost and biogas slurry are, therefore, contradictory. I argue that, firstly, the stronger immediate effect displayed with CaSa-compost fertilisation is the possible result of a direct rise in soil pH through simultaneous liming. Soil pH, however, is not a parameter in SWIM, even though it is highly relevant for predicting nutrient availability in the soil. Secondly, there is the possibility that SWIM overestimates N fertilisation and underestimates P fertilisation. With respect to beans, I found that a seasonal biomass growth of beans in fresh matter of at least 30 t/ha is needed to reach the break-even threshold where the balance of natural input and output flows of N turns from a net negative to a net positive result (Krause and Rotter 2018). This corresponds to a crop yield of about 3.8 t/ha of air-dried beans (ibid.). This yield has only been possible with the use of CaSa-compost as a fertiliser (Krause et al. 2016).

5   Discussion of Opportunities and Challenges

This section summarises important lessons learned with respect to real-world applications of the technologies analysed and the recycling fertilisation practices studied. Potential bottlenecks, and opportunities for how to overcome them, are identified regarding the use of biogas slurry, biochar and human excreta as, or in, recycling fertilisers.

5.1 Utilising Biogas Slurry as a Fertiliser

I identified the following potential opportunities and challenges for the practical use of biogas slurry in smallholder farming.

Biogas Slurry is an Adequate Fertiliser, but Not an 'Untapped Resource' in All Cases

Locally available biogas slurry is characterised by nutrient contents (Table in Krause et al. 2016) that are adequate for fertilisation, as compared to the literature (e.g. Finck 2007; Horn et al. 2010). Ample quantities of such biogas slurry are available for those smallholders operating a biogas digester (Krause and Rotter 2017). However, these material flows should not be considered as an 'untapped resource' for fertilising msiri fields, as both input materials for biogas fermentation have previously been used as fertiliser input in shambas, as shown by Baijukya and de Steenhuijsen Piters (1998). It is thus vital that biogas slurry is recycled into banana-based home gardens in order to replace prior inputs of banana stem and cow dung and to avoid an exacerbation of existing nutrient depletion in shamba systems (ibid.).

Integrating Biogas Slurry in Soil Management May Create 'Islands of Soil Fertility'

Farmers that already possess a BiogaST digester perform slurry management as follows. Some slurry is directly removed from the digester with buckets and used as a fertiliser. Further slurry leaves the digester through an outlet hole and flows via a small runlet into a pit filled with grasses and/or cow dung. After pre-composting in the pit, slurry is used to fertilise tomatoes in kitchen gardens, maize grown in msiri fields or banana plants in shambas. The latter practice is highly recommended, especially when banana stems are used as fermentation substrate (cf. prior paragraph). This practice potentially creates 'islands of soil fertility' in the vicinity of farm houses and, at the same time, accelerates existing 'discrete patterns of soil fertility', as described by Tittonell (2016). According to practical experience, a significant share of biogas slurry remains in the pit as the total amount available is too much for farmers to manage. This means, however, that the fertiliser effect of slurry is untapped, whilst greenhouse gas (GHG) emissions and nutrient leaching both increase.

Fertilisation Benefits of Biogas Slurry Are Minor When Compared to the Effect of Compost

When experimenting with biogas slurry as a soil amender, bean and maize plants did not respond as well to biogas slurry as to compost or to CaSa-compost (Krause et al. 2016). In addition, MAVUNO has observed that tomato plants developed bigger plants but rather small fruits when fertilised with biogas slurry in demonstration plots. Also, Komakech et al. (2015) did not find a specific advantage to using fermented matter compared to composted matter when studying biomass growth and crop yields of maize plants in Uganda. One possible explanation is that the comparatively high levels of NH_4^+ and organic acids in biogas slurry are phytotoxic for plants (cf. Möller and Müller 2012; Salminen et al. 2001). Composting of biogas slurry with other organic matter could, therefore, reduce these phytotoxic substances, as demonstrated by Abdullahi et al. (2008). Farmers may directly exchange their practical experiences of using either liquid or composted biogas slurry for growing various local crops when meeting in 'biogas clubs'.

Biogas Slurry Allows for a 'Target-Oriented' Fertiliser Application

A specific opportunity and supposed advantage of biogas slurry, compared to compost, is the possibility to synchronise nutrient applications with crops' nutrient demands (Möller and Müller 2012). This 'target-oriented' fertiliser application is particularly relevant for crops with increased N demand (ibid.). Taking maize as an example, plant nutrient requirements are at their highest levels between 28 and 42 days after sowing (Kuratorium für Technik und Bauwesen in der Landwirtschaft (KTBL) 2009). Applying biogas slurry is, therefore, most highly recommended during this maturing stage.

Net Effects on Gaseous Emissions After Slurry Application Are Not yet Quantifiable

When using biogas slurry as a soil amender, N_2O contributes significantly to GHG emissions from the agroecosystem, whilst N leaching and NH_3 contribute marginally to eutrophication (Krause and Rotter 2018). Incorporating biogas slurry into the soil shortly (i.e. during the 12 h) after application potentially avoids N_2O and NH_3 emissions, pursuant to Möller et al. (2008) and Möller and Stinner (2009). Farmers may use a hand-hoe to cover the slurry, or apply the slurry directly into a hole/furrow, which they then fill in immediately after. Furthermore, biogas slurry should preferably be applied to dry soil (ibid.). Amon et al. (2006) recommend a high water content of the slurry, or additional dilution of the slurry with water, to allow for rapid infiltration. The water content of local biogas slurry is approximately 95% of the fresh matter (Krause et al. 2015) and should thus be adequate for this purpose. When comparing the environmental impacts of using biogas slurry, compost or mulch, existing literature is ambiguous. In stockless organic cropping systems, for example, more than one-third of N_2O emissions could be avoided by digesting crop residues before reallocation to fields, compared to using the same residues for mulching (Möller and Stinner 2009). In organic cropping systems with animal manuring, NH_3 and N_2O emissions and leaching of NO_3^- after application of biogas slurry are, meanwhile, comparable to those associated with applications of compost or mulching (Möller 2015). Overall, the major effect on environmental emissions after using biogas slurry, compost or mulch can be summarised as follows: on the one hand, high levels of N in biogas slurry from co-fermentation of cow dung increase N_2O emissions as nitrification is promoted (Möller and Stinner 2009). On the other hand, comparatively low levels of organic C in biogas slurry potentially decrease N_2O emissions as the activity of C decomposers (bacteria and fungi) is inhibited. This, in turn, reduces the availability of NH_4^+ for nitrification and, thus, denitrification (Möller 2015). Overall, according to Möller (2015), field applications of biogas slurry primarily affect soil microbial activities, not GHG emissions.

5.2   Utilising Biochar for Composting

I identified the following potential opportunities and challenges in regard to the practical use of biochar for composting.

Total Recovery Potential of Biochar from Households Is Sufficient for CaSa-Composting

Smallholder households cooking with microgasifiers potentially 'produce' biochar in fresh matter of 270–300 kg/year, respectively, depending on whether a sawdust gasifier or a TLUD is used (Krause and Rotter 2017). Finally, total potential to recover (maximum) biochar, on the one hand, and sanitised solids, on the other, results in comparable annual material flows in terms of volume (ibid.). This, in turn, fits very well into the practices required to produce CaSa-compost (Krause et al. 2015). After composting, about 2.6–2.8 m^3 of CaSa-compost is available to smallholders each year (Krause and Rotter 2018). This amount is generally adequate and feasible for handling, carrying and amending compost in the soil (Sanchez et al. 1997). Furthermore, I add one personal experience from my scientific experiments described in Krause et al. (2016). During the field trials, I observed that CaSa-compost aided workability of the soil, by making it more friable, presumably due to the biochar content. However, I did not make any scientific analysis to follow up on this anecdotal observation.

Biochar Promotes Liming and, Thus, Improves the Efficiency of Nutrient Recycling

Nutrient availability in the soil is, amongst other factors, an outcome of soil pH (Horn et al. 2010). The optimal topsoil pH range for cropping is 5.5–6.5 (ibid.). To buffer acids in soils, and thus to neutralise soil acidity, common measures include the use of lime ($CaCO_3$) (ibid.) and/or the addition of organic material (e.g. Wong et al. 1998). The addition of biochar is also associated with soil liming (cf. Biederman and Harpole 2013; Jeffery et al. 2011; Liu et al. 2013). In this respect, I found that biochar-containing CaSa-compost is characterised by a higher liming potential (per kg of N added to the soil) than biogas slurry, standard compost or other organic or synthetic fertilisers (Krause et al. 2015). This theoretically assessed liming potential is further practically effective in significantly increasing the soil pH to 6.1 within one season, compared with just 5.3 on unamended soil (Krause et al. 2016). In comparison, highly productive Terra Preta soils, which are particularly rich in biochar, are characterised by a pH of 5.2–6.4, comparatively higher than surrounding soils, as shown by Glaser and Birk (2012). The annual production of CaSa-compost is sufficient for application over a total area of >1000 m^2 per year and to maintain, or even improve, soil pH sustainably within this area (Krause and Rotter 2018). As a consequence, liming improves nutrient availability in the soil and renders nutrient cycling through organic residues more effective.

The Effects of Composting Biochar on Environmental Emissions Are Not yet Quantifiable

Emissions from composting contribute significantly to overall environmental emissions from the agroecosystem (Krause and Rotter 2018). In particular, CO_2 and N_2O add to the global warming potential, whilst NH_3 and P leaching contribute to eutrophication (ibid.). The fact that biochar captures NO_3^- and PO_4^{3-}, as shown, for example, by Agyarko-Mintah et al. (2016), Gronwald et al. (2015) and Kammann et al. (2015), is promising in regard to reducing GHG emissions and nutrient leaching during composting and also after compost is added to the soil. There is also empirical evidence that adding biochar to composting can decrease N_2O and CH_4 emissions (e.g. Agyarko-Mintah et al. 2016; Sonoki et al. 2013; Vandecasteele et al. 2013; Wang et al. 2014). When co-composting biochar with urine, as is practised in CaSa-composting, NH_3 emissions still rise, but the increase observed is lower than after solely adding urine to compost (Larsen and Horneber 2015). Similarly, N_2O and CH_4 emissions also decrease, when adding urine and biochar to compost, whilst those emissions increase when only urine is added (ibid.). Results from observing changes in soil-borne emissions after using biochar are more ambiguous. According to Cayuela et al. (2014) and Zhang et al. (2012), N2O emissions from biochar amended soils are lower than those from unamended soils. Additionally, net fluxes of CH_4 from managed soils decrease after biochar amendment (ibid.). This depends, however, on soil moisture levels and oxygenation (Van Zwieten et al. 2015). In contrast, other researches have found increased emissions of N_2O (Singh et al. 2010), as well as of CO_2 and CH_4 (e.g. Spokas et al. 2009; Zhang et al. 2012), after amending biochar to soils. Overall, existing scientific data still exposes uncertainties in various areas and knowledge gaps in relation to the underlying principles and mechanisms at play (cf. Mukherjee and Lal 2014; Van Zwieten et al. 2015). It remains, therefore, a challenge to quantify changes in net emissions from an agroecosystem which utilises biochar for composting and soil fertility management.

5.3   Utilising Faeces as a Compost Additive

I identified the following potential opportunities and challenges for the practical use of faeces for composting.

Human Faeces Contribute Significantly to the Nutrient Content of Compost, Especially P

An important aspect of CaSa-compost is its comparatively high nutrient content (Krause et al. 2015). Contents of P_{tot} and N_{min}, for example, are about three times higher in CaSa-compost than in Karagwe standard compost (ibid.). My findings are supported by considerable Ptot and Nmin concentrations in composts blended with faeces, as compared to compost without the addition of faeces, as per another experiment I took part in (Krause and Klomfaß 2015). Given that faeces are characterised by significantly higher P content than, for example, grass cuttings or kitchen waste (cf. Krause and Rotter 2018), the remarkably high P content in CaSa-compost can be attributed to co-composted human faeces. My model-based analysis shows that human faeces contribute about one-quarter of the P contained in CaSa-compost (ibid.). In view of the fact that P scarcity is a major soil constraint threatening farmers in Karagwe, this is a highly significant and beneficial aspect of utilising human faeces in agriculture.

Faecal Compost Increases Crop Growth, but Its Use Is Not Recommended for All Crops

In the field experiment, I tested the viability of CaSa-compost to effectively increase biomass growth and yields for maize, beans, cabbage, onion and carrots (Krause et al. 2016). The benefits observed for CaSa-compost, which uses locally available nutrients, match those of significantly higher inputs of synthetic N fertiliser. Likewise, Andreev et al. (2016) have observed a higher increase in maize grain yields when using faecal-biochar compost, in comparison to synthetic fertilisers or animal manure, in Moldova. Consequently, maize plants grown in soil amended with faecal matter-blended compost take up significantly larger amounts of P and N than plants grown using standard compost (Krause et al. 2016; Krause and Klomfaß 2015). Adequate concentrations of those macronutrients can, therefore, be found in plant tissue and seeds (ibid.). Meanwhile, composts which contain human faeces should, in general, not be used for crops which grow underground, such as for onion, carrots, beetroot, potatoes, etc. (e.g. Richert et al. 2010). Nonetheless, as shown in the field trial, CaSa-compost was especially beneficial for cultivating maize and beans (Krause et al. 2016). The demo plots of MAVUNO further indicate the appropriateness of CaSa-compost for growing tomatoes.

Human Excreta Can Alternatively Be Used for Growing Perennial Crops or for Reforestation

If the use of human excreta for the production of annual crops is undesirable for farmers, there are other options for its agricultural use. For example, faecal compost could be applied to fields used for growing bamboo, fodder grasses such as elephant grass (*Pennisetum purpureum*, *Miscanthus fuscus* or *Miscanthus violaceus*) or other grasses, which can be used as a fodder or energy crop. Another option for utilising human excreta is for growing bananas in shambas. This practice is called omushote in Swahili and was common in Karagwe until pit latrines were installed in the region through 'development cooperations' in the 1940s (Rugalema et al. 1994). It would be possible to adapt this method to modern practice in the following way. Faeces are first collected in a double-vault UDDT and then pre-composted inside the toilet for a period lasting from several weeks to months. Pre-composted solids are then applied, on rotational basis, to planting holes for banana plant cuttings. Similarly, faeces may be applied to the planting holes for trees, including fruit trees and trees for use as firewood or timber. To avoid transmission of diseases through direct contact or through flies, the hole needs to be covered ultimately with a layer of soil of about 30–50 cm in depth. Such reforestation could, for example, be realised on remote fields in the vicinity of settlements and thereby contribute to ameliorating degraded soils in these areas.

Mixing Human Faeces with Other Organic Matter Ensures Adequate Composting

However, when faeces are employed for composting, they should always be mixed with other kinds of organic residues, such as kitchen waste, harvest residues and also biochar or ashes. The aim of this is to sustain a well-functioning composting process with a balanced mixture of (1) C- and nutrient-rich material, (2) fractions of easily degradable organics and of stable matter suitable for humification and (3) dry and wet matter (e.g. Amlinger et al. 2008; Heinonen-Tanski and van Wijk-Sijbesma 2005; Niwagaba et al. 2009).

5.4   Utilising Urine as Fertiliser and Compost Additive

I identified the following potential opportunities and challenges for the practical use of urine in agriculture.

Different Means of Treating Urine Prior to Application to Reduce Odour

It is likely that the most common method of using urine is as a liquid fertiliser diluted with water (Richert et al.2010). Urine is diluted mainly (1) to avoid excessive application of urine and (2) to reduce odour. If urine is used undiluted, Richert et al. recommend applying it to a furrow or hole and closing the furrow/hole with soil thereafter. This can reduce odour and N losses through sub-surface volatilisation. Lactic acid fermentation of urine is another option to reduce odour (e.g. Andreev et al. 2017). During the fermentation process, the lactic acid produced inhibits urease and, thus, also inhibits the formation of ammonia, whilst still conserving urea (ibid.). In practice, the lactic acid bacterial inoculum should be added to the empty urine storage tank of the UDDT before urine collection starts (ibid.). This increases the efficiency of lactic acid fermentation. Another approach to using urine as a fertiliser is the addition of magnesium oxide to stored urine, which results in a crystalline product called 'struvite', or magnesium ammonium phosphate (MAP) (Winker et al. 2011).

Fertilising with Urine, in General, Promotes Plant Growth

The beneficial effects of using urine as fertiliser have often been demonstrated (e.g. Andersson 2015; Esrey et al. 2001; Richert et al. 2010; Schönning and Stenström 2004). Arnold and Schmidt (2012) found that both treatments, stored urine and struvite, show equally good fertilising characteristics and are thus valid substitutes to synthetic fertiliser for cultivating maize, beans, summer wheat or *Miscanthus*. When testing the use of urine specifically for Karagwe (Krause et al. 2016), the urine's qualities as a fertiliser were altered inside the UDDT by passing it through a deodoriser block in the urinal. For this reason, the true benefit of urine fertilisation was not easy to gauge in our experiment.

Adding Urine Maintains a Well-Functioning Composting Process

The benefits of adding urine to compost identified in this study are as follows: (1) urine contributes to moisture levels in the mixture; (2) urine enriches the compost product with N and P; and (3) adding urine to compost can reduce workload for farmers. In regard to the first point, practical experience from the CaSa case study showed that adding urine to the compost pit was highly effective in order to avoid the matter drying out in local climate conditions in Karagwe. As water is scarce in the region, utilising urine solves this problem without the need for extra (fresh) water. To maintain a well-functioning composting process and to minimise NH_3 and N_2O emissions from the process, Amlinger et al. (2008) recommend a stable moisture content in the mixture of between 50% and 60% of the total fresh matter. In regard to the second point, to maintain a fast and odourless process and also to minimise GHG emissions, C and N in the compost mixture should be in the range of 25–35 (i.e. C/N ratio) (ibid.). The optimal C/N range is determined by the needs of microorganisms, which require proportional content of C as an energy carrier and N for proteins (Finck 2007). Most materials added to compost tend to be rather rich in C, and so the C/N ratio of the resulting compost is often too high (ibid.), which slows down the process. According to my model, the C/N ratio in Karagwe standard compost is about 43 and in CaSa-compost about 38 (Krause and Rotter 2018). Hence, the C/N ratio in CaSa-compost is lower and therefore more proportional, presumably due to N input from urine. In regard to the third point, using urine-blended compost can avoid an additional work step, when compared to applying compost prior to sowing or planting and then adding urine separately during plant growth.

Enhancing Compost with Urine Potentially Decreases the Efficiency of N Recovery

With regard to the overall efficiency of N recovery, I found that, theoretically, the N-recovery potential is higher for the direct field application of urine than for its use as a compost additive, with approximately 70% and 55% of total N recovered, respectively (Krause and Rotter 2018). From a practitioner's perspective, there is, thus, a trade-off between optimising the efficiency of N recovery, improving the composting process and managing or reducing workload. Finally, since biochar can capture NO_3^- and PO_4^{3-} (cf. Sect.), the combined use of urine and biochar for composting also has the potential to reduce nutrient loss during, and after, compost-ing and thus to positively affect the turnover of N and P and to make the practice of blending compost with urine more effective.

Factors Surrounding Pharmaceuticals and Hormones in Urine Are, as yet, Largely Unstudied

Finally, there are existing challenges in relation to 'health and hygiene' that relate to unknowns and uncertainties about 'organic micropollutants' (OMPs) contained in human excreta, such as pharmaceuticals, hormones, etc. Most OMPs are contained in urine, whilst faeces contain the larger part of pathogens (Richert et al. 2010). The eventual fate of these hazardous substances and the risk they pose to local populations is an important issue, which is, however, neither especially relevant to Karagwe nor specifically related to EcoSan. To the best of my knowledge, it is still unclear on a global scale how we will deal sustainably with different OMPs that we continuously and increasingly emit into the ecosystem. OMPs are also released into the environment through water toilets and pit latrines (e.g. Ngumba et al. 2016), most often in an uncontrolled manner and without any further treatment. Similarly, little is known about methods to eliminate OMPs in the environment, the prevalence of OMPs in the soil, their uptake by plants, further consequences for human health through consumption, the eco-toxicity of metabolisms, etc. It is certain, however, that simply storing urine is not sufficient to completely remove pharmaceuticals from urine. Struvite, meanwhile, is a product free of pharmaceuticals and pathogens (Schürmann et al. 2012). Furthermore, according to the WHO (2006), 'the soil system is generally better equipped than watercourses for the degradation of pharmaceutical residues'. Further to this point, Arnold (2012) observed the successful degradation of hormones in the soil. With respect to pharmaceuticals, Arnold and Schmidt (2012) demonstrated that diclofenac (painkiller), atenolol (beta blocker) and verapamil (high blood pressure treatment) are not transferred to crops when present in urine used as fertiliser. Carbamazepine (an epilepsy treatment), however, can be determined in maize grains and stalks, but in relatively low concentrations (ibid.). Furthermore, nitrification is a suitable process for removing certain pharmaceuticals, whilst filtration of urine with activated carbon is a viable way to remove all pharmaceuticals from urine (Bischel et al. 2015). However, secure disposal of used activated carbon needs to be solved.

6   Conclusions

In order to be considered 'sustainable', soil fertility management should, amongst other factors, mitigate existing soil constraints, such as nutrient depletion and soil acidity. In the contemporary context, it should also promote resilience for agriculture in the face of climate change as a local and global threat. SOM, and the restoration of SOM, is of equal importance to climate adaptation measures, as it contributes to the soil's water-holding capacity and erosion resistance. Ultimately, soil fertility management is applied in order to maintain or improve crop productivity. With respect to the potential identified for capturing residues from cooking and EcoSan for recycling-driven soil fertility management, I conclude that recovering and processing residues from smallholder households provides a significant opportunity to increase access to fertiliser and soil improvers through subsistence production. All treatments analysed are viable as substitutes for synthetic, commercial fertilisers, but CaSa-compost displays benefits over and above the alternatives. The potential of CaSa-compost for 'sustainable' soil fertility management is superior to that of standard compost, especially with respect to liming and potential SOM restoration. Biogas slurry gives inferior results in all aspects when compared to compost amendments, but especially for liming, potential SOM restoration and GHG emissions. Moreover, even when the strong P-retention characteristics of the local Andosol are taken into consideration, further gradual increases in soil P are possible with regular applications of CaSa-compost. Both prevailing challenges for agricultural production in Karagwe – namely, P scarcity and soil acidification – can be mitigated through sufficient application rates of CaSa-compost, as the analysed case studies showed. Whether, and how, CaSa practice can also serve as a mitigation measure in relation to climate change could be the focus of future research. The demonstrated potential to increase yields is theoretically sufficient to enable food sovereignty for smallholder households, whilst the corresponding nutrient requirements are adequately compensated for by locally available residual matter. Therefore, this practical approach of recovering biochar and human excreta for soil fertility management represents an exit strategy from the vicious circle of poor soil quality and insufficient production of food crops and residual matter in the context of smallholdings in sub-Saharan Africa. Overall, my results endorse the establishment of a clear link between cooking, sanitation and agriculture and, therefore, an 'intersectional resource management' approach around the energy-sanitation-agriculture nexus. This is due to the fact that recyclable matter from energy and sanitation facilities has complementary benefits. For example, biochar from microgasifiers promotes the recycling of C for restoring SOM, whilst those residues collected from EcoSan facilities contribute to capturing nutrients for fertilisation.

References

Abdullahi, Y. A., Akunna, J. C., White, N. A., Hallett, P. D., & Wheatley, R. (2008). Investigating the effects of anaerobic and aerobic post-treatment on quality and stability of organic fraction of municipal solid waste as soil amendment. *Bioresource Technology, 99*(18), 8631–8636.

Agyarko-Mintah, E., Cowie, A., Singh, B. P., Joseph, S., Van Zwieten, L., Cowie, A., Harden, S., & Smillie, R. (2016). Biochar increases nitrogen retention and lowers greenhouse gas emissions when added to composting poultry litter. *Waste Management, 61*, 138–143.

Amlinger, F., Peyr, S., & Cuhls, C. (2008). Green house gas emissions from composting and mechanical biological treatment. *Waste Manage Resources, 26*(1), 47–60.

Amon, B., Kryvoruchko, V., Amon, T., & Zechmeister-Boltenstern, S. (2006). Methane, nitrous oxide and ammonia emissions during storage and after application of dairy cattle slurry and influence of slurry treatment. *Agriculture, Ecosystems and Environment, 112*(2), 153–162.

Andersson, E. (2015). Turning waste into value: Using human urine to enrich soils for sustainable food production in Uganda. *Journal of Clean Production, 96*, 290–298.

Andreev, N., Ronteltap, M., Lens, P. N., Boincean, B., Bulat, L., & Zubcov, E. (2016). Lacto-fermented mix of faeces and bio-waste supplemented by biochar improves the growth and yield of corn (Zea mays L.). *Agriculture, Ecosystems & Environment, 232*, 263–272.

Andreev, N., Ronteltap, M., Boincean, B., Wernli, M., Zubcov, E., Bagrin, N., Borodin, N., & Lens, P. N. (2017). Lactic acid fermentation of human urine to improve its fertilizing value and reduce odour emissions. *Journal of Environmental Management, 198*, 63–69.

Arnold, U. (2012). Landwirtschaftliche Nutzung von Gelbwasser und MAP –praktischer Einsatz und rechtliche Rahmenbedingungen. *Presentation at the conference 'New water infrastructure concepts for urban planning'*. Eschborn, Germany.

Arnold, U., & Schmidt, J. (2012). Research Project Sanitary Recycling Eschborn (SANIRESCH), Project component: Agricultural Production/Legal Situation. *Deutsche Gesellschaft für Internationale Zusammenarbeit (GIZ)*. Eschborn, Germany.

Baijukya, F. P., & de Steenhuijsen Piters, B. (1998). Nutrient balances and their consequences in the banana-based land use systems of Bukoba district, northwest Tanzania. *Agriculture, Ecosystems & Environment, 71*, 147–158.

Baijukya, F. P., De Ridder, N., & Giller, K. E. (2006). Nitrogen release from decomposing residues of leguminous cover crops and their effect on maize yield on depleted soils of Bukoba District, Tanzania. *Plant and Soil, 279*(1–2), 77–93.

Batjes, N. H. (2011). *Global distribution of soil phosphorus retention potential*. ISRIC Report 2011/06, Wageningen University and Research Centre, Wageningen, Netherlands, 41.

Batjes, N. H., & Sombroek, W. G. (1997). Possibilities for carbon sequestration in tropical and subtropical soils. *Global Change Biology, 3*(2), 161–173.

Biederman, L. A., & Harpole, W. S. (2013). Biochar and its effects on plant productivity and nutrient cycling – A meta-analysis. *GCB Bioenergy, 5*(2), 202–214.

Bischel, H. N., Özel Duygan, B. D., Strande, L., McArdell, C. S., Udert, K. M., & Kohn, T. (2015). Pathogens and pharmaceuticals in source-separated urine in eThekwini, South Africa. *Water Research, 85*, 57–65.

Buresh, R. J., Smithson, P. C., & Hellums, D. T. (1997). Building soil phosphorus capital in Africa. In *Replenishing soil fertility in Africa* (Special Publication 51) (pp. 111–149). Chicago: Soil Science Society of America and American Society of Agronomy (SSSA).

CaSa. (2012). *Project report from Engineers Without Borders for the pilot phase of the project "Carbonization and sanitation" in Karagwe, Tanzania*.

Cayuela, M. L., Van Zwieten, L., Singh, B. P., Jeffery, S., Roig, A., & Sánchez-Monedero, M. A. (2014). Biochar's role in mitigating soil nitrous oxide emissions: A review and meta-analysis. *Agriculture, Ecosystems and Environment, 191*, 5–16.

Chesworth, W. (Ed.). (2008). *Encyclopedia of soil science* (p. 902). Dordrecht: Springer.

Christensen, T. H., Gentil, E., Boldrin, A., Larsen, A., Weidema, B., & Hauschild, M. (2009). C balance, carbon dioxide emissions and global warming potentials. *Waste Manage Resources, 27*(8), 707–715.

DeSchutter, O. (2011). Agroecology and the right to food. *Report by the special rapporteur on the right to food of the United Nations (UN), presented at the 16th session of the UN Human Rights Council [A/HRC/16/49]*, Geneva, Switzerland.

dePauw, E. D. (1984). *Soils, physiography and agro-ecological zones of Tanzania*. Crop. Monitoring and Early Warning Systems Project. GCPS/URT/047/NET, Kilimo/FAO, Dar es Salaam.

Esrey, S. A., Andersson, I., Hillers, A., & Sawyer, R. (2001). *Closing the Loop – Ecological sanitation for food security*. Publications on Water Resources No. 18. Swedish International Development Cooperation Agency (SIDA) Sweden/Mexico.

Falcão, N. P. S., Clement, C. R., Tsai, S. M., & Comerford, N. B. (2009). Pedology, fertility, and biology of central Amazonian Dark Earths. In *Amazonian dark earths: Wim Sombroek's vision* (pp. 213–228). Dordrecht: Springer, Netherlands.

FAO. (2014). A statement by FAO director-general José Graziano da Silva. In *24th session of the Committee on Agriculture (COAG) opening statement*. Rome: Food and agriculture Organization of the United Nations (FAO).

FAOSTAT. (2012). *Statistics of the FAO, annual, data for: Maize; yield; all countries*. Statistical Databases of the FAO.

Finck, A. (2007). *Pflanzenernährung und Düngung in Stichworten (Plant nutrition and fertilization in keywords)* (6th ed.). Stuttgart: Borntraeger.

Glaser, B., & Birk, J. J. (2012). State of the scientific knowledge on properties and genesis of anthropogenic dark earths in Central Amazonia (terra preta de Índio). *Geochima est Cosmochima Acta, 82*, 39–51.

Gornott, C., Hattermann, F., & Wechsung, F. (2015). Yield gap analysis for Tanzania – The impact of climate and management on Maize yields. In *Poster presentation at 'Management of land use systems for enhanced food security – Conflicts, controversies and resolutions'*. Berlin: Tropentag.

Gornott, C., Hattermann, F., & Wechsung, F. (2016). Liebig's law – Increase and stabilise Tanzanian Maize yields by combining different crop modelling approaches. In *Oral presentation at 'Solidarity in a competing world – Fair use of resources*. Vienna: Tropentag.

Graeub, B. E., Chappell, M. J., Wittman, H., Ledermann, S., Kerr, R. B., & Gemmill-Herren, B. (2016). The state of family farms in the world. *World Development, 85*, 1–15.

Gronwald, M., Don, A., Tiemeyer, B., & Helfrich, M. (2015). Effects of fresh and aged biochars from pyrolysis and hydrothermal carbonization on nutrient sorption in agricultural soils. *Soil Discussions, 2*, 29–65.

Heinonen-Tanski, H., & van Wijk-Sijbesma, C. (2005). Human excreta for plant production. *Bioresource Technology, 96*(4), 403–411.

Horn, R., Brümmer, G. W., Kandeler, E., Kögel-Knabner, I., Kretzschmar, R., Stahr, K., & Wilke, B. M. (2010). *Scheffer/Schachtschabel – Lehrbuch der Bodenkunde (Textbook of soil science)* (16th ed., p. 570). Heidelberg: Springer Spektrum.

Jeffery, S., Verheijen, F. G. A., van der Velde, M., & Bastos, A. C. (2011). A quantitative review of the effects of biochar application to soils on crop productivity using meta-analysis. *Agriculture, Ecosystems and Environment, 144*(1), 175–187.

Kammann, C., Schmidt, H. P., Messerschmidt, N., Linsel, S., Steffens, D., Müller, C., Koyro, H. W., Conte, P., & Stephen, J. (2015). Plant growth improvement mediated by nitrate capture in co-composted biochar. *Nature Scientific Reports, 5*(11080), 12.

Kenneth, F. G. Masuki, J., Mbogoni, G., & Ley, J. (2003). Agro-ecological zones of the lake zone, Tanzania. In *Lake Zone Agricultural Research and Development Institute (LZARDI) and National Soil Service, Agricultural Research Institute (ARI) Mlingano* (p. 26). Tanzania.

Kimaro, A. A., Timmer, V. R., Chamshama, S. A. O., Ngaga, Y. N., & Kimaro, D. A. (2009). Competition between maize and pigeonpea in semi-arid Tanzania: Effect on yields and nutrition of crops. *Agriculture, Ecosystems and Environment, 134*, 115–125.

Kimetu, J. M., Lehmann, J., Ngoze, S. O., Mugendi, D. N., Kinyangi, J. M., Riha, S., & Pell, A. N. (2008). Reversibility of soil productivity decline with organic matter of differing quality along a degradation gradient. *Ecosystems, 11*(5), 726–739.

Komakech, A. J., Zurbrügg, C., Semakula, D., Kiggundu, N., & Vinnerås, B. (2015). Evaluation of the performance of different organic fertilisers on maize yield: A case study of Kampala, Uganda. *Journal of Agricultural Science, 7*(11).

Krause, A. (2018). *Valuing wastes – An integrated system analysis of bioenergy, ecological sanitation, and soil fertility management in smallholder farming in Karagwe, Tanzania*. Dissertation at TU Berlin, Germany.

Krause, A., & Klomfaß, J. (2015). Kohlenstoff- und Nährstoffrecycling mit Bioenergie- und ökologischen Sanitärsystemen. In *Presentation at the workshop 'Biokohle im Gartenbau – Verwertung von organischen Reststoffen zur Schließung von Energie- und Stoffkreisläufe'*. Berlin: Botanical Garden.

Krause, A., & Rotter, V. S. (2017). Linking energy-sanitation-agriculture: Intersectional resource management in smallholder households in Tanzania. *Science of the Total Environment, 59*, 514–530.

Krause, A., & Rotter, V. S. (2018). Recycling improves soil fertility management in smallholdings in Tanzania. *Agriculture, 8*(3), 31.

Krause, A., Kaupenjohann, M., George, E., & Koeppel, J. (2015). Nutrient recycling from sanitation and energy systems to the agroecosystem – Ecological research on case studies in Karagwe, Tanzania. *African Journal of Agricultural Research, 10*(43), 4039–4052.

Krause, A., Nehls, T., George, E., & Kaupenjohann, M. (2016). Organic wastes from bioenergy and ecological sanitation as a soil fertility improver: A field experiment on a tropical Andosol. *The Soil, 2*, 147–162.

Krysanova, V., Wechsung, F., Arnold, J., Srinivasan, R., & Williams, J. (2000). *SWIM (Soil and Water Integrated Model) user manual*. PIK Report Nr. 69, p. 239.

KTBL. (2009). Faustzahlen für die Landwirtschaft (Rule-of-thumb figures for agriculture). *Kuratorium für Technik und Bauwesen in der Landwirtschaft (KTBL)* (14th edn.). Darmstadt, Germany.

La Via Campesina. (2015). Declaration of the International Forum for Agroecology.

Lal, R. (2006). Managing soils for feeding a global population of 10 billion. *Journal of the Science of Food and Agriculture, 86*(14), 2273–2284.

Lal, R. (2009). Soils and world food security. *Soil & Tillage Research, 102*(1), 1–4.

Landon, J. R. (1991). *Booker tropical soil manual: a handbook for soil survey and agricultural land evaluation in the tropics and subtropics.* 1st paperback edition (p. 474). Essex: Longman Scientific & Technical Ltd.

Larsen, O., & Horneber, D. (2015). Einfluss von Biokohle auf Treibhausgasemissionen während der Kompostierung.In K. Terytze (Ed.), *Book of abstracts of the scientific workshop "Biokohle im Gartenbau – Verwertung von organischen Reststoffen zur Schließung von Energie- und Stoffkreisläufen"* (pp. 33–34). Berlin, Germany.

Lehmann, J., & Joseph, S. (2009). *Biochar for environmental management – Science and technology.* Sterling: Earthscan.

Liu, J., Schulz, H., Brandl, S., Miehtke, H., Huwe, B., & Glaser, B. (2012). Short-term effect of biochar and compost on soil fertility and water status of a dystric Cambisol in NE Germany under field conditions. *Journal of Plant Nutrition and Soil Science, 175*, 698–707.

Liu, X., Zhang, A., Ji, C., Joseph, S., Bian, R., Li, L., & Paz-Ferreiro, J. (2013). Biochar's effect on crop productivity and the dependence on experimental conditions—A meta-analysis of literature data. *Plant and Soil, 373*(1–2), 583–594.

Markwei, C., Ndlovu, L., Robinson, E., & Shah, W. P. (2008). *International Assessment of Agricultural Knowledge, Science and Technology for Development (IAASTD): Sub-Saharan Africa summary for decision makers.* Washington, DC: Island Press.

McIntyre, B. D., Herren, H. R., Wakhungu, J., & Watson, R. T. (2009). Agriculture at a crossroads. In *IAASTD: Global report.* Washington, DC: Island Press.

Möller, K. (2015). Effects of anaerobic digestion on soil carbon and nitrogen turnover, N emissions, and soil biological activity. A review. *Agronomy for Sustainable Development, 35*(3), 1021–1041.

Möller, K., & Müller, T. (2012). Effects of anaerobic digestion on digestate nutrient availability and crop growth: A review. *Engineering in Life Sciences, 12*(3), 242–257.

Möller, K., & Stinner, W. (2009). Effects of different manuring systems with and without biogas digestion on soil mineral nitrogen content and on gaseous nitrogen losses (ammonia, nitrous oxides). *European Journal of Agronomy, 30*(1), 1–16.

Möller, K., Stinner, W., Deuker, A., & Leithold, G. (2008). Effects of different manuring systems with and without biogas digestion on nitrogen cycle and crop yield in mixed organic dairy farming systems. *Nutrient Cycling in Agroecosystems, 82*(3), 209–232.

Mourice, S. K., Rweyemamu, C. L., Tumbo, S. D., & Amuri, N. (2014). Maize cultivar specific parameters for decision support system for agrotechnology transfer (DSSAT) application in Tanzania. *American Journal of Plant Sciences, 5*, 821–833.

Mukherjee, A., & Lal, R. (2014). The biochar dilemma. *Soil Resources, 52*, 217–230.

Ngumba, E., Gachanja, A., & Tuhkanen, T. (2016). Occurrence of selected antibiotics and antiretroviral drugs in Nairobi River Basin, Kenya. *Science of the Total Environment, 539*, 206–213.

Niwagaba, C., Nalubega, M., Vinnerås, B., Sundberg, C., & Jönsson, H. (2009). Bench-scale composting of source-separated human faeces for sanitation. *Waste Management, 29*(2), 585–589.

Nziguheba, G. (2001). *Improving phosphorus availability and maize production through organic and inorganic amendments in phosphorus deficient soils in western Kenya.* Doctoral thesis, dissertationes de agricultura no. 462, Katholieke Universiteit Leuven, Belgium.

Ogola, S. A. (2013). *Land and natural resources conflict in transboundary agroecosystem management project Kagera basin.* Assessment report to project 'Kagera Agro-Ecosystems'. FAO, Rome, Italy.

Peel, M. C., Finlayson, B. L., & McMahon, T. A. (2007). Updated world map of the Köppen-Geiger climate classification. *Hydrology and Earth System Sciences Discussions, 4*(2), 439–473.

Richert, A., Gensch, R., Jönsson, H., Stenström, T-A., & Dagerskog, L. (2010). *Practical guidance on the use of urine in crop production.* Stockholm Environment Institute (SEI), EcoSanRes Programme, Stockholm.

Rugalema, G. H., Okting'Ati, A., & Johnsen, F. H. (1994). The homegarden agroforestry system of Bukoba district, North-Western Tanzania. 1. Farming system analysis. *Agroforestry Systems, 26*(1), 53–64.

Salminen, E., Rintala, J., Härkönen, J., Kuitunen, M., Högmander, H., & Oikari, A. (2001). Anaerobically digested poultry slaughterhouse wastes as fertiliser in agriculture. *Bioresource Technology, 78*(1), 81–88.

Sanchez, P. A., Shepherd, K. D., Soule, M. J., Place, F. M., Buresh, R. J., Izac, A. M. N., Mokwunye, U., Kwesiga, F. R., Ndiritu, C. G., & Woomer, P. L. (1997). Soil fertility replenishment in Africa: An investment in natural resource capital. Replenishing soil fertility in Africa. *Soil Science Society of America, 51*, 1–46.

Schönning, C., & Stenström, A. T. (2004). Guidelines for the safe use urine and faeces in ecological sanitation systems. In *Swedish Institute for infectious disease control, EcoSanRes Programme.* Stockhoml: SEI.

Schrecker, S. (2014). *Konstruktionsplanung eines Brenners zur Nutzung von Biogas als Kochwärme und experimentelle Effizienzbestimmung für die lokale Implementierung in Tansania.* Diploma/ Master thesis, TU Berlin, Germany.

Schürmann, B., Everding, W., Montag, D., & Pinnekamp, J. (2012). Fate of pharmaceuticals and bacteria in stored urine during precipitation and drying of struvite. *Water Science and Technology, 65*(10), 1774–1780.

Singh, B. P., Hatton, B. J., Singh, B., Cowie, A. L., & Kathuria, A. (2010). Influence of biochars on nitrous oxide emission and nitrogen leaching from two contrasting soils. *Journal of Environmental Quality, 39*, 1224–1235.

Sonoki, T., Furukawa, T., Jindo, K., Suto, K., Aoyama, M., & Sánchez-Monedero, M. Á. (2013). Influence of biochar addition on methane metabolism during thermophilic phase of composting. *Journal of Basic Microbiology, 53*(7), 617–621.

Spokas, K. A., Koskinen, W. C., Baker, J. M., & Reicosky, D. C. (2009). Impacts of woodchip biochar additions on greenhouse gas production and sorption/degradation of two herbicides in a Minnesota soil. *Chemosphere, 7*, 574–581.

Springmann, M., Clark, M., Mason-D'Croz, D., Wiebe, K., Bodirsky, B. L., Lassaletta, L., De Vries, W., Vermeulen, S. J., Herrero, M., Carlson, K. M., et al. (2018). Options for keeping the food system within environmental limits. *Nature, 562*(7728), 519.

Tittonell, P. (2016). Feeding the world with soil science: Embracing sustainability, complexity and uncertainty. *Soil Discussions,* in review.

UNEP. (2007). *Reactive nitrogen in the environment: Too much or too little of a good thing.* Paris: United Nations Environment Programme (UNEP).

URoT. (2012). National sample census of agriculture 2007/2008.*Regional report – Kagera region, volume Vh.* United Republic of Tanzania, Ministry of Agriculture, Food Security and Cooperatives, The National Bureau of Statistics and the Office of the Chief Government Statistician, Zanzibar, Tanzania.

Van Zwieten, L., Kammann, C., Cayuela, M., Singh, B. P., Joseph, S., Kimber, S., Donne, S., Clough, T., & Spokas, K. A. (2015). Biochar effects on nitrous oxide and methane emissions from soil. In *Biochar for environmental management* (pp. 489–520). London: Routledge.

Vandecasteele, B., Mondini, C., D'Hose, T., Russo, S., Sinicco, T., & Quero Alba, A. (2013). Effect of biochar amendment during composting and compost storage on greenhouse gas emissions, N losses and P availability. In *Proceedings of 15th RAMIRAN international conference, recycling of organic residues in agriculture.* Versailles, France.

Wang, Y., Dong, H., Zhu, Z., Li, T., Mei, K., & Xin, H. (2014). Ammonia and greenhouse gas emissions from biogas digester effluent stored at different depths. *Transactions of the ASABE, 57*(5), 1483–1491.

WHO. (2006). *WHO guidelines for the safe use of wastewater, excreta and greywater – Volume 4. Excreta and greywater use in agriculture.* Geneva: World Health Organization (WHO), WHO Press.

Willett, W., Rockström, J., Loken, B., Springmann, M., Lang, T., Vermeulen, S., et al. (2019). Food in the Anthropocene: The EAT–lancet commission on healthy diets from sustainable food systems. *The Lancet, 393*(10170), 447–492.

Winker, M., Paris, S., Heynemann, J., & Montag, D. (2011). Phosphorrückgewinnung aus Urin mittels Struvitfällung in einem Frankfurter Bürogebäude. *fbr-wasserspiegel, 1*(11), 3–4.

Wong, M. T. F., Nortcliff, S., & Swift, R. S. (1998). Method for determining the acid ameliorating capacity of plant residue compost, urban waste compost, farmyard manure, and peat applied to tropical soils. *Communications in Soil Science & Plant Analysis, 29*(19–20), 2927–2937.

Zakharova, A., Beare, M. H., Cieraad, E., Curtin, D., Turnbull, M. H., & Millard, P. (2015). Factors controlling labile soil organic matter vulnerability to loss following disturbance as assessed by measurement of soil-respired $\delta^{13}CO_2$. *European Journal of Soil Science, 66*(1), 135–144.

Zhang, A., Liu, Y., Pan, G., Hussain, Q., Li, L., Zheng, J., & Zhang, X. (2012). Effect of biochar amendment on maize yield and greenhouse gas emissions from a soil organic carbon poor calcareous loamy soil from Central China Plain. *Plant and Soil, 351*, 263–275.

Traditional and Adapted Composting Practices Applied in Smallholder Banana-Coffee-Based Farming Systems: Case Studies from Kagera and Morogoro Regions, Tanzania

Anika Reetsch, Didas Kimaro, Karl-Heinz Feger, and Kai Schwärzel

Abstract In Tanzania, about 90% of the banana-coffee-based farming systems lie in the hands of smallholder farmer families. In these systems, smallholder farmers traditionally add farm waste to crop fields, making soils rich in organic matter (humus) and plant-available nutrients. Correspondingly, soils remained fertile during cultivation for over a century. Since the 1960s, the increasing demand for food and biofuels of a growing population has resulted in an overuse of these farming systems, which has occurred in tandem with deforestation, omitted fallows, declined farm size, and soil erosion. Hence, humus and nutrient contents in soils have decreased and soils gradually degraded. Inadequate use of farm waste has led to a further reduction in soil fertility, as less organic material is added to the soils for nutrient supply than is removed during harvesting. Acknowledging that the traditional use of farm waste successfully built up soil fertility over a century and has been reduced in only a few decades, we argue that traditional composting practices can play a key role in rebuilding soil fertility, if such practices are adapted to face

A. Reetsch (✉)
Institute for Integrated Management of Material Fluxes and of Resources (UNU-FLORES), Dresden, Germany

Sebastian Kolowa Memorial University, Lushoto, Tanzania
e-mail: reetsch@unu.edu

D. Kimaro
Sebastian Kolowa Memorial University, Lushoto, Tanzania

K.-H. Feger
Technische Universität Dresden, Dresden, Germany
e-mail: karl-heinz.feger@tu-dresden.de

K. Schwärzel
Thünen Institute of Forest Systems, Eberswalde, Germany
c-mail: kai.schwaerzel@thuenen.de

the modern challenges. In this chapter, we discuss two cases in Tanzania: one on the traditional use of compost in the Kagera region (Great African Rift Valley) and another about adapted practices to produce compost manure in the Morogoro region (Uluguru Mountains). Both cases refer to rainfed, smallholder banana-coffee-based farming systems. To conclude, optimised composting practices enable the replenishment of soil nutrients, increase the capacity of soils to store plant-available nutrients and water and thus, enhance soil fertility and food production in degraded banana-coffee-based farming systems. We further conclude that future research is needed on a) nutrient cycling in farms implementing different composting practices and on b) socio-economic analyses of farm households that do not successfully restore soil fertility through composting.

Keywords African smallholder agriculture · Banana-coffee-based farming systems · Reuse of farm waste · Composting · Soil fertility and conservation

1 Introduction

Increasing yield gaps for almost all crops and decreasing food security are exacerbating poverty in rural areas in Sub-Saharan countries (Tittonell and Giller 2013). In many cases, low biomass production results from soil and land degradation, which in turn are driven or accelerated by three factors: firstly, the growing demand for food of an increasing population; secondly, poor soil and land management; and thirdly, the increasing variability of rainfall pattern due to climate change (Masawe 1992; FAO 2017a; Gebrechorkos et al. 2018).

In banana-coffee-based farming systems in mountainous regions in Tanzania, smallholder farming contributes up to 95% of agricultural production (cf. FAO 2017a). Mountainous regions in Tanzania are densely populated and intensively used to produce banana (mainly fruit banana and plantain, *Musa* spp.), coffee (mainly *Coffea caneophora*), maize, roots, tubers, pulses, and legumes. Since the 1960s, agricultural production in these regions has increased, often in unsustainable ways. For instance, omitted fallows, intensive use of woody biomass in three-stone fires, missing awareness concerning soil erosion measures, frequent tillage, and lack of composting and mulching have resulted in the exploitation of vegetation, land, and soils. In rural areas of Tanzania, such as the Kagera region, more than 90% of the rural households' cooking energy relies on firewood (unprocessed woody biomass) and charcoal (processed woody biomass), and improved stoves are not widely utilised (URT 2012). Due to the poor soil and land management, the amount of organic material added to the soils has reduced compared to the soil status in the 1960s and 1990s (Copeland Reining 1967; Touber and Kanani 1996). This led to a reduction of humus content and plant-available soil nutrients and thus, decreasing agricultural production. Since the extraction of biomass is not compensated by measures to improve soil fertility and nutrient recycling, significant degradation of vegetation and soil, and thus, accelerated soil nutrient depletion and declining water resources occur in the mid and long terms (Schwärzel et al. 2017). Furthermore, East African countries have experienced increased unreliability in rainfall patterns in the last decade, and future scenarios show that higher temperatures, less rain, and changes in rain pattern are very likely for some regions in Tanzania (Gebrechorkos et al. 2018). Smallholder farmers in mountainous regions in Tanzania, however, depend on rainfed agriculture, and we experienced that only a few farm households have experience with or the capacity of long-term water harvesting to feed the family's demand for drinking water and to irrigate the main farmland. Food production is therefore severely jeopardised.

Traditionally, smallholder farming practices in banana-coffee-based farming systems ensured that a sufficient amount of organic material returns to the soils to produce a thick, dark-coloured, humus-rich, and thus, nutrient-rich topsoil (Copeland Reining 1967; Masawe 1992; Touber and Kanani 1996). As introduced above, unsustainable agricultural practices since the 1960s have led to a continuous reduction in the addition of organic material to the topsoil, and thus, to a reduction of humus content in the soil. With less humus that is able to store and release nutrients, the productivity of these farming systems has diminished. A change in reuse and recycling of organic material from farms is thus needed. Research has shown that the potential of farm waste as a soil fertiliser and soil conditioner is not yet exhausted in banana-coffee-based farming systems because adapted composting practices are not frequently applied (cf. Kimaro et al. 2011). Furthermore, nutrients contained in human excreta are not widely maximised (Krause et al. 2015, 2016). Acknowledging that the traditional use of farm waste successfully built up soil fertility over at least one century and has been reduced in only a few decades, we argue that traditional composting practices will have to play a key role in rebuilding soil fertility in degraded banana-coffee-based farming systems. To transform degraded banana-coffee-based farming systems into sustainable agroforestry or agroecology systems that meet regional and global challenges, the understanding of traditional uses of farm waste is as important as the integration and adaptation of these practices (Gliessman 2015, 2016; FAO 2015, 2017b).

In this chapter, we focus on traditional and adapted composting practices in banana-coffee-based farming systems to highlight the positive properties of organic farm waste as a soil fertiliser and soil conditioner. In the following, we first describe the characteristics of banana-coffee-based farming systems, and secondly, illustrate traditional and adapted composting practices that are typical for these farming systems. Then, we present two cases where these composting practices are applied. The first case introduces the work of the farmer initiative Mavuno Project in the Kagera region in north-west Tanzania (Great African Rift Valley, Lake Victoria Basin; Fig.. In the Mavuno Project, about 750 smallholder farm families have been trained in implementing adapted composting practices to restore degraded banana- coffee- based farming systems. The second case presents the work of a farmer field school established by the Sokoine University of Agriculture (SUA) in the Uluguru Mountains, in the Morogoro region in central Tanzania (Kimaro et al. 2011). Among the skills trained, farmers learnt how to produce adapted in-situ and on-surface composting. In the discussion, we compare degraded banana-coffee-based farming

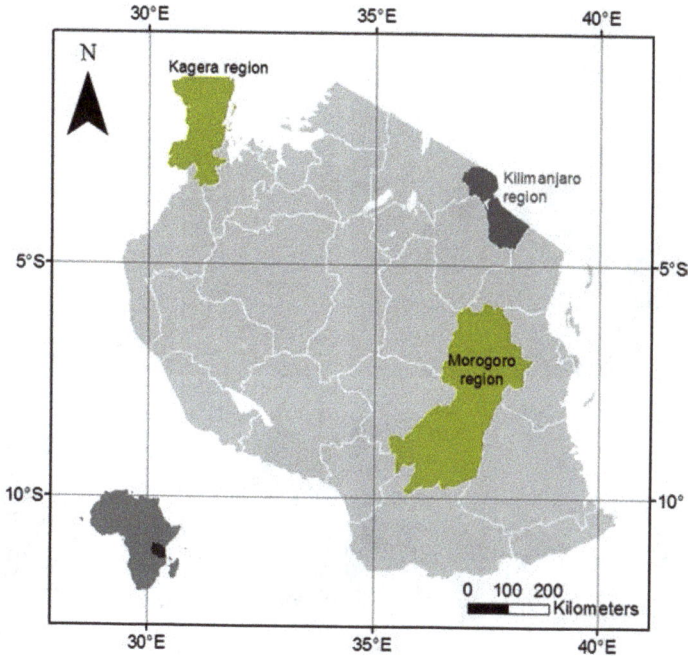

Fig. Map of the Kagera, Kilimanjaro, and Morogoro regions in Tanzania. (Vector files used from Map Library 2007)

systems with densely cropped and well-managed systems in the Kilimanjaro region (north Tanzania), discuss the advantages and disadvantages of the composting practices introduced, and consider their limitations.

2 Banana-Coffee-Based Farming Systems in the Highlands of Tanzania

The banana-coffee-based farming system is a typical smallholder, usually rainfed, subsistence farming system based on agroforestry in a tropical, mountainous environment, covering the dominant perennial crops coffee and banana, several annual crops, and native trees (Rugalema et al. 1994). Garrity et al. (2012) classified it as a typical Sub-Saharan African farming system, namely, the 'highland perennial', indicating that perennial crops—in this case, banana and coffee—are the core of the agricultural production. In Tanzania, traditional banana-coffee-based farming systems are mostly cultivated by smallholder farmer families and consist of up to four subsystems: the homegarden called *Kibanja* (in other Bantu languages named '*Kihamba*', '*Shamba*', or '*Chagga*'), new farmland or land in transition from

Fig. The subsystems of a banana-coffee-based farming system in the Kagera region, north-west Tanzania: (**a**). *Kibanja*, (**b**) *Kikamba*, (**c**) *Rweya*, (**d**) *Kabira*

grassland to farmland called the *Kikamba*, grassland called the *Rweya*, and woodland, the *Kabira* (Copeland Reining 1967; Rugalema et al. 1994; Baijukya et al. 2005; Hemp and Hemp 2008; Dancer 2015; Fig.).

Depending on the region, the naming of these subsystems differs according to the local Bantu language. In the Kilimanjaro region, the banana–coffee-based farming system is known as the *Chagga* (Hemp and Hemp 2008), named after the dominant tribe, the Chagga, that settled on the slopes of the Mount Kilimanjaro. Chagga landowners create and maintain densely intercropped and productive homegardens known as *Kihamba* where cultivation is well established. Around the *Kihamba* a small area is set aside for women to grow a variety of vegetables, which include amaranth, cabbage, peas, and tomatoes. A multilayered vegetation structure, which corresponds to that of a tropical mountain forest with trees, shrubs, and herbs can be found within the *Kihamba* (Akinnifesi et al. 2008). Comparable to the *Kihamba* in the Kilimanjaro region, the *Kibanja* in the Kagera region is the heart of the farming system. In these fields, which usually are the closest to the

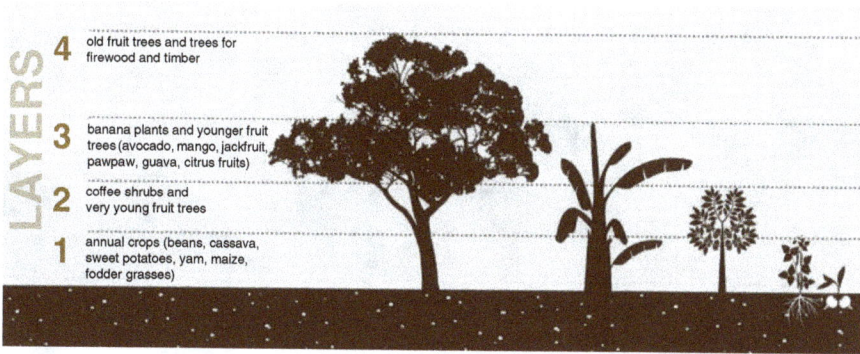

Fig. Multilayered vegetation in the traditional banana-coffee-based farming system—the *Kibanja* or *Kihamba*—in Tanzania. (Based on Rugalema et al. 1994 and Akinnifesi et al. 2008, Design: Claudia Matthias)

farmers' homes, biomass production is the highest and thus secures the livelihood of the farm families (Copeland Reining 1967; Rugalema et al. 1994).

In very densely cropped homegardens under favourable soil conditions, the vegetation corresponds to that of a tropical mountain forest and consists of multiple layers: annual crops (first layer), coffee shrubs and very young fruit trees (second layer), banana plants and younger fruit trees (third layer), and older trees (fourth layer) (Copeland Reining 1967; Rugalema et al. 1994; Akinnifesi et al. 2008; Hemp and Hemp 2008; Fig.). The first layer is up to 1 m high. Here, beneath a canopy of coffee bushes and banana plants, a variety of shade-tolerant annual food crops grow, such as beans, cassava, maize, yams, sweet and Irish potato, and also fodder, herbs, and grass. Coffee, medicinal plants and shrubs, and a few species of young trees are found within the second canopy zone, which lies approximately between a height of 1 and 2.5 m. Less commonly cultivated are the perennial crops vanilla, cotton, and sugarcane. The third vegetation layer, with the banana canopy along with other kinds of fruit and fodder trees, is located above 2.5 m and approximately reaches a height of 5 m. Here, various banana varieties are grown, of which plantain, the cooking banana is the primary staple food for the farm households; the sweet finger banana is cultivated as a fruit; and the brewing banana to brew local beer. Above this, the fourth layer is less distinct and more blended together. It contains various kinds of trees delivering shade and fruit crops, for example avocado, mango, pawpaw, jackfruit, and citrus fruits, as well as fodder, timber, and firewood. The shade provided by the trees plays an important role in reducing soil evaporation. The response of bananas to droughts is complex; drought effects are associated with low yields notably 6–8 months afterwards. Besides, stall-fed livestock activities and the cultivation of vegetables in kitchen gardens are common practices in this farming system. The farmer families keep mainly goats, pigs, sheep, and chicken in the homegardens, and in a few cases, improved cattle for milk production.

Fig. Livestock rearing in smallholder farming: (**a**) traditional, free-range indigenous cattle in the *Rweya*, (**b**) integrated livestock management with cowshed for improved cattle in the *Kibanja*, (**c**) tied goats under a roof in the *Kibanja*, (**d**) free-range pigs in the *Kibanja*

As local weather conditions vary, several combinations of water-harvesting practices can be considered in agricultural production. In the Kagera region, for example, some farmers gather rainwater in clay containers from the roofs of their houses to irrigate the kitchen gardens, whereas a series of interlinked canals carrying water harvested in the forest on Mount Kilimanjaro, called *mfongo*, helps irrigate the homegardens of the *Chagga* people in the Kilimanjaro region. There, the canals deliver a convenient source of water for domestic use as well as for irrigation purposes. As coffee and banana require between 900 and 1050 m^3 water ha^{-1} month^{-1} (Baulme 1993), irrigation water is supplied via the canals at intervals of 5–8 weeks, in the case of insufficient rainfall.

The other three subsystems are less fertile and diverse. The *Kikamba* refers to land transitioning from *Rweya* towards *Kibanja* (cf. Baijukya 2004). The original vegetation is gradually replaced by young banana seedlings and some annual crops. Very often, the soils are uncovered and particularly exposed to soil erosion. The *Rweya* is grassland and shrubland, which are often used for free-range livestock-rearing. Traditionally, livestock rearing is an essential strategy against food shortage in dry seasons and an additional source of income (Lichtfield and McGregor 2008). The fourth subsystem is the *Kabira*, a land parcel with trees for firewood, charcoal, and timber production (Copeland Reining 1967).

3 Composting Practices

In the banana-coffee-based farming system, the following kinds of organic farm waste are produced: crop residues, livestock urine and manure, kitchen and food waste, litter, dead wood, ashes (inorganic), animal bones, and human urine and faeces; however, not all kinds of farm waste are used to produce compost, particularly not human excreta (Krause et al. 2015, 2016). Since open defaecation is prohibited, human excreta are collected in pit latrines, which are not sealed at the bottom and from which the excreta can easily leak into the underlying groundwater aquifer. Human excreta or pit latrine sludge is not redistributed to the fields. The subsystems of a banana-coffee-based farming system play an essential role in the composting process because farm waste and biomass (grass and wood)—and thus, plant nutrients—circulate within the *Kibanja* and also between *Kibanja*, *Kikamba*, *Rweya*, and *Kabira*.

In general, in the *Kibanja*, most parts of the farm waste usually return to the crop fields, except human excreta, which is gathered in pit latrines on the *Kibanja* land

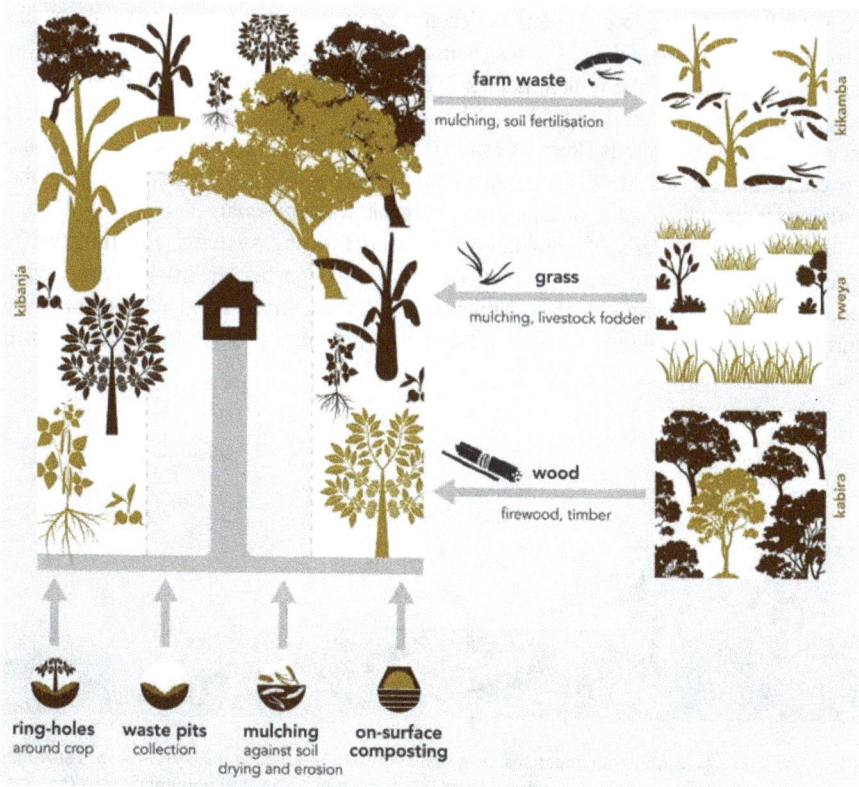

Fig. Flows of significant farm waste, grass, and firewood in traditional banana–coffee-based farming systems. (Based on Baijukya and Steenhuijsen Piters 1998), Design: Claudia Matthias

and is not reused. To fertilise the *Kikamba* land and to protect its topsoil against erosion, farmers remove organic material originating from the *Kibanja* and add it to the soils of the *Kikamba* land. From the *Rweya*, grass and firewood are imported to the *Kibanja*. The grass is either used as fodder or as mulch. Leftovers of the burnt firewood usually stay in the *Kibanja*. Wooden biomass is imported from the *Kabira* into the *Kibanja* and leftovers of the use of wooden biomass usually stay in the *Kibanja*.

3.1 Traditional Practice: In-Situ and Pit Composting

In the *Kibanja*, different practices of composting have been developed over the centu-ries and are still applied by most of the smallholder farmers (Reetsch et al. 2020a). The primary practices of traditional composting are ring-hole composting, in-situ compost-ing, and pit composting (Below Fig.). As indicated in Below Fig. a, farmers dig ring-holes around perennial crops, fill them with nutrient-rich farm waste, preferably livestock manure, and cover the filled plots with soil material (*ibid.*). Another way of composting is presented in Below Fig. b. In-situ composting comprises of farm waste, which remains in place or can be spread over the surface and left to decompose itself, without any amend-ments like layering, watering, or adding of ashes (Kimaro et al. 2011; Reetsch et al. 2020a). During the rotting process of the mulch, humus is accumulated in the topsoil. A third form of reusing farm waste is pit composting (Below Fig. c). Very often, farmers col-lect different kinds of farm waste in pits (Reetsch et al. 2020a). After filling these waste pits, they are covered with soil and, if available, grass (*ibid.*). Then, a new pit is dug. Over time, the farmland is spread with pits containing rotten organic waste material.

In the homegardens of banana-coffee-based farming systems, the farm waste is composted either in situ or in pits (Below Fig.). In particular, in-situ composting enables the replenishment of soil nutrients and the conservation of soil moisture throughout the farmland, whereas pit and ring-hole composting improves the soil nutrient status at selected sites.

Fig. Traditional composting practices in banana–coffee-based farming systems in Tanzania: (**a**) in-situ composting: ring-hole application of livestock manure around perennial crops (banana, coffee), (**b**) in-situ composting: crop field mulched with plant-based farm waste, (**c**) pit compost-ing: farm waste mixed in a waste pit (Design: Claudia Matthias)

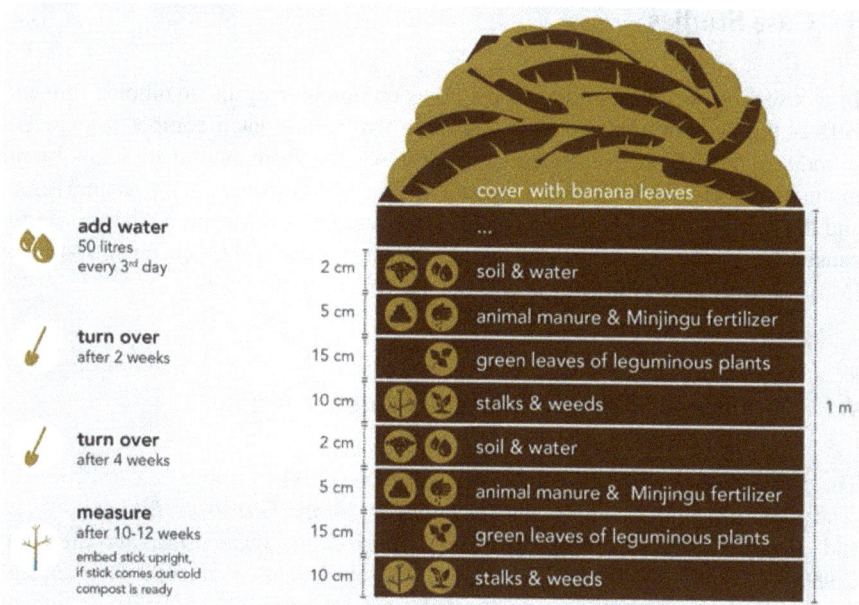

Fig. The production of on-surface compost in smallholder farmers in banana-coffee-based farming systems in Tanzania according to Kimaro et al. (2011). (Design: Claudia Matthias)

3.2 Adapted Practice: On-Surface Composting

Traditional composting practices have been modified and further developed as on-surface composting (heap method) by Kimaro et al. (2011). This aboveground technique facilitates the production of compost, especially by women and children. The compost pile consists of several layers of farm waste, which are piled on top of each other above the ground. In the first step, the ground is covered with a plastic sheet to avoid leaching, followed by an approximately 10 cm thick layer of stalks and weed, which forms the base of the compost pile. The second layer con-sists of green leaves of leguminous plants (15 cm in height) to enrich the compost with nitrogen. The third layer preferably contains animal manure and the local, phosphorus-rich Minjingu rock fertiliser (5 cm in height) followed by a 2 cm soil layer that is moistened with water. The Minjungu rock is mined in the north of Tanzania. All four steps are repeated until a height of approximately 1 m is reached. Finally, the compost pile is covered with banana leaves to control evaporation. To promote the rotting process, the compost pile needs to be regularly turned over. During the rotting process, a wooden stick is pierced through the layers to estimate the core temperature of the compost pile. As long as the rotting process is ongoing, the stick is warm and moist. If it becomes cold and dry, the rotting process is completed.

4   Case Studies

In this section, we present two case studies on composting in smallholder banana-coffee- based farming systems in Tanzania that follow each composting practice introduced in the previous section. Both cases are from humid to semi- humid mountainous regions with steep slopes. The first case is located in the Kagera region and the second in the Morogoro region. Both regions suffer from land degra-dation caused by human activities and climate change as outlined in the introduction (Masawe 1992; FAO 2017a).

4.1   Traditional Composting in the Kagera Region

The Kagera region lies in north-west Tanzania 1,200 m above sea level (a.s.l.). The predominant soils are Rhodic and (*Anthri-*) *Humic Ferralsols, Lithic* and *Mollic Leptosols, Humic Acrisols, Anthri-luvic Phaeozems,* and *Ferralic Cambisols* following the FAO-UNESCO soil classification of 1988 (Touber and Kanani 1996) and recently as *Andosols* by Krause et al. (2016). With an annual rainfall ranging between 800 and 1,000 mm, falling in two rainy seasons, which allows for two cropping seasons per year. However, changing rainfall patterns were reported to cause harvest shocks in the last two decades (Trærup and Mertz 2011). The soils in the Kagera region degraded due to increased rainfall variability, soil nutrient losses, and deforestation—and thus increased soil erosion (Baijukya and Steenhuijsen Piters 1998; Trærup and Mertz 2011; Wasige et al. 2013; FAO 2017b)—and changed since 1901 from tropical forest, woodland, and savanna, to cropland and pasture (Wasige et al. 2013).

Traditionally, the use of farm waste plays an important role in this region (Copeland Reining 1967; Katoke 1970; Ndege et al. 1995; Touber and Kanani 1996). Over at least one century, farmers continuously added organic plant material and livestock manure to the fields in the homegardens (Copeland Reining 1967). The warm-humid climate conditions of the region are favourable to decompose organic material within a few days or weeks. This promotes the formation of A-horizons in the soil with a thickness of 30 to 40 cm, rich in soil organic matter and plant-available nutrients (Touber and Kanani 1996). Nutrient levels are found to be especially high near the farmers' houses with a decreasing gradient to the borders of the farm (Baijukya and Steenhuijsen Piters 1998). The thickness of the A-horizons decreased over the last five decades because the fields have been intensively used for agricultural production to meet the increasing demand for food and biofuels of the fast-growing and refugee-hosting population (CARE and Overseas 1994; URT 2016, 2018;).

The farmer initiative Mavuno Project (https://mavunoproject.or.tz/wp/) has sup-ported smallholder farmer families in restoring degraded traditional banana-coffee-farming systems and further transforming the *Kibanja* into sustainably intensified

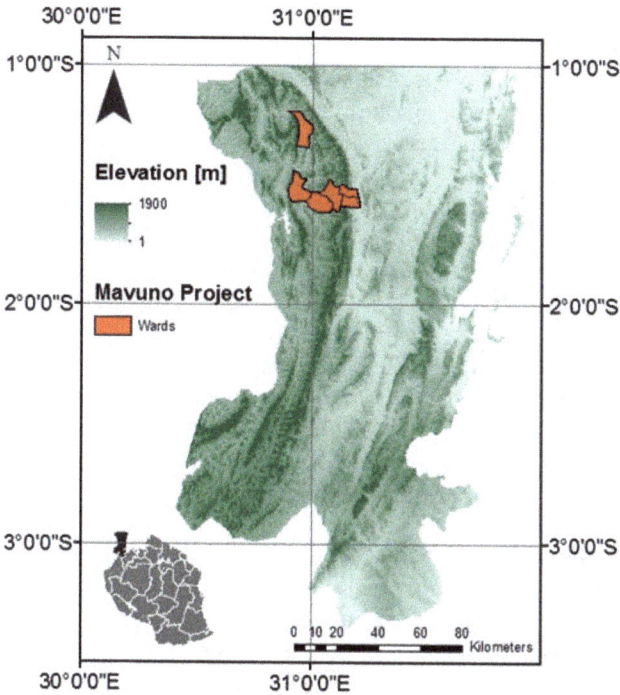

Fig. Map of the Kagera region in north-west Tanzania with elevation contours. (Digital Elevation Model used from CGIAR-CSI (2017))

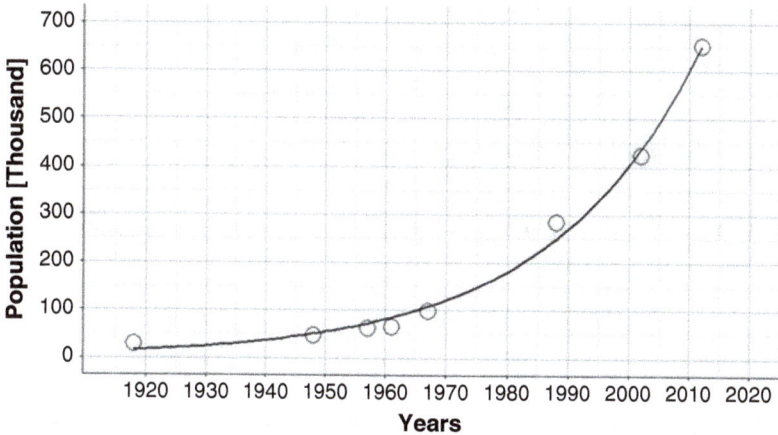

Fig. Population growth in Karagwe and Kyerwa districts, Kagera region. (Data extracted from Copeland and Reining (1967), Katoke (1970), and URT (2016))

and climate-resistant farming systems. The agricultural transformation was jointly developed with smallholder farmers acknowledging traditional composting practices, involving climate change adaptation measures, and land management practices that are also known in agroforestry (Wekesa and Jönsson 2014; Gliessman 2016; FAO 2017a). These measures involved nutrient management, soil and water conservation, the establishment of mixed crop-tree systems, adapted agricultural practices (crop rotation, intercropping, relay and contour strip cropping), tillage, residue management, integrated livestock management, sustainable energy use, and integrated pest management (Wekesa and Jönsson 2014).

Inspired by this smallholder farmer initiative, we investigated the implementation of traditional and adapted composting practices in five focus group discussions with 22 lead farmers working with the Mavuno Project (Reetsch et al. 2020b). The lead farmers were responsible for training and monitoring of 750 smallholder farm families in the Kyerwa and Karagwe districts in reintegrating traditional composting practices and optimising waste flows within the farmland.
According to the focus groups, the farm households began by establish-ing a plot of one acre (0.4 ha) where they planted perennial and annual crops inter-cropped with a few trees. In this early stage, optimised in-situ composting was already practised. With every rainy season, the plot got more diverse, and farmers started to establish further plots. In the discussions, the lead farmers indicated that soil fertility and biomass production increased in those farm households that implemented the composting practices that they were trained on, whereas they did not increase when a farm household did not properly apply the practices taught. As a side effect, humus enrichment counteracted with soil acidification, protected soil water from evaporation, and markedly reduced soil erosion.
Increasing soil fertility led to higher yields and higher food availability for the entire farm household. Harvested crops could be stored either for consumption or for sale. However, the first results also indicated that not all farm households succeed in the same way.

4.2 On-Surface Composting in the Morogoro Region

In the Uluguru Mountains in the Morogoro region, the mean annual rainfall varies with altitude, from 900 mm at 550 m a.s.l. to 2,300 mm at more than 1,500 m a.s.l. In this area, mountain ridges are mainly used for the production of banana, vegetable, bean, and short rain maize while on the foothills long rain maize is the main crop (Kimaro et al. 2005). The topography is highly variable with dissected mountain ridges and foothills with very steep slopes of up to 80% and narrow valleys (Kimaro et al. 1999). Bedrock geology is dominated by metasediments mainly consisting of hornblende pyroxene granulite, with plagioclase and quartz-rich veins (Kimaro et al. 2005). Based on the FAO (2006), the soils on the mountain ridges are dominantly *Endoskeletic* and *Leptic Cambisols*, with accessory surfaces of *Haptic* and *Chromic Phaeozems* and *Orthi-eutric Regosols*. On the foothills, the dominant soils are *Chromic Lixisols* and *Profondic Acrisols* associated with *Hyperferralic*

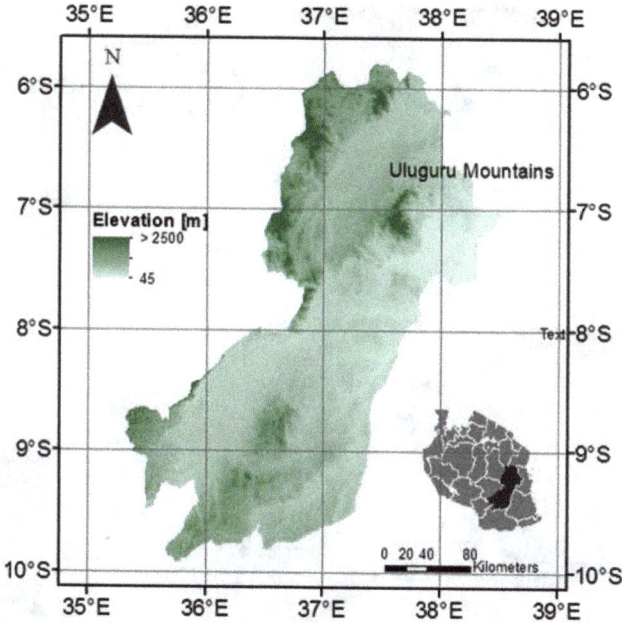

Fig. Map of the Morogoro region in central East Tanzania. (Digital Elevation Model used from CGIAR-CSI (2017))

Cambisols and *Endoleptic Cambisols*. The soil covers are generally affected by severe erosion. Increased free-range livestock and human activity have led to the collapse of the soil conservation system and increased land degradation (including soil erosion). Therefore, strategies that aimed at combating land degradation through mechanical and biological measures such as reforestation activities, agroforestry, protection of watersheds, improved land husbandry, and environmental conservation were initiated in Tanzania in the late 1980s (Shetto 1999).

The Sokoine University of Agriculture (SUA) initiated a soil and water conservation programme in the Uluguru Mountains to train smallholder farmers on the effect of conservational tillage on soil loss and plant nutrient status on their fields. To that end, a farmer field school was piloted in the north-eastern part (Kimaro et al. 2005). The farmers were trained on how to increase the productivity of their farming systems through composting. This included the application of in-situ and on-surface composting in combination with conservation tillage and terracing. The results showed that yields under in-situ composting and conservation tillage were the highest, presumably due to improved soil fertility and decreased soil losses in space and time (Kimaro et al. 2011, Fig.).

Seasonal yields of vegetable were measured in order to assess the following soil and water conservation practices: (1) conservational tillage and in-situ compositing, (2) traditional terrace and in-situ compositing, (3) conservational tillage and manure, and (4) traditional terrace and manure over traditional farming practices. In this

Fig. Training farmers on improved indigenous farming practices: (**a**) and (**c**) On-surface composting, (**b**) and (**d**) Traditional terraces (Kimaro et al. 2011)

case, adapted in-situ composting means a mixture of green manure (a mixture of *Gliricidia* and other farm residues at a rate of 5 t ha^{-1}), farmyard manure (10 t ha^{-1}), and Minjingu phosphate rock (MPR) 100 kg P ha^{-1} were left to decompose in situ on the soil (Msita et al. 2010). The results showed that conservation tillage with in-situ composting followed by traditional terrace and in-situ composting gave a higher fresh yield of vegetables, presumably due to improved soil fertility and decreased soil losses in space and time, than conservation tillage and livestock manure alone, traditional terrace and manure alone, controlled conservation tillage, and controlled traditional terrace (Msita et al. 2010). It can be concluded for this case that integrating composting practices in conservation tillage for crop production on sloping land is the best practice for sustainable crop production, nutrient availability, and reduction of soil loss.

5 Discussion

As evidenced in both cases, adding compost materials to soils enhances the content of soil organic matter and soil nutrients in degraded banana-coffee-based farming systems. The elevated humus levels increased soil fertility and caused an increase in yield. In the first case, in the Kagera region, previously productive banana-coffee-based farming systems had deteriorated due to rapid land and soil degradation. The desperate attempt to feed and supply the people (locals and refugees from Rwanda and Burundi) with firewood in the 1990s accelerated the loss of biodiversity, thinned vegetation density and the thick, humus-enriched topsoil. However, through reintegration and optimisation of local traditional composting practices, farmers were able to restore soil fertility and thus, to increase biomass production. In our opinion, degraded banana-coffee-based farming systems could reach a high level of diversity through composting. In contrast, in the banana-coffee-based farming systems of the Kilimanjaro region, however, soil and land degradation had been successfully counteracted by composting for several decades (Hemp and Hemp 2008; Kimaro et al. 2011). There, biodiversity had been highly maintained with a densely grown vegetation structure of four layers and thus agricultural production systems have remained multifunctional (food, biofuels, cash crops, medicine) (cf. Figure). Degraded homegardens, as presented in the first case in the Kagera region, were less densely cropped with fewer plant species and consisted of only one or two, seldom three vegetation layers (cf. Reetsch et al. 2020a, Figure). In our opinion, multilayered farming systems in the Kilimanjaro region can be understood as a target state for degraded banana-coffee-based farming systems in the Kagera region.

In both cases, we also observed the challenges and limitations of composting practices. Although composting is often seen as an easily realisable technique, five principles have to be considered to achieve the intended effects (Kimaro et al. 2011; Wekesa and Jönsson 2014; FAO 2015). First, not every farming system produces or

Fig. Densely grown *Kihamba* in the Kilimanjaro region

needs the same kind of compost. In fact, it depends on the biomass produced in the farming system. Second, the materials to produce compost often compete with other uses, for example, with mulching or livestock feeding. Third, climate/weather conditions strongly control the rotting process through moisture and temperature. If it is too dry or too cold, microorganisms may not properly decompose and transform the organic raw materials. Fourth, collecting farm waste and caring for an ongoing rotting process, which includes several turnovers of the decomposing material, and the application of compost to the crop fields is time and labour-intensive. Fifth, the composition and quantity of plant residues differ among crops and, when used in composting, the nutrient composition of the resulting compost material changes as a result. Besides, the main disadvantage of traditional pit composting is that soils are only punctually enriched with organic matter. The compost material stays in the pit and is not distributed to the crops. As these traditional pits consist mainly of kitchen waste, they are often installed near the farmhouses. Nutrient concentrations are thereby higher in the vicinity of the farmhouses and decrease with increasing distance. However, only numerous rotted waste pits between the crops would bring the soil nutrients to the roots of the crops. Compared to in-situ and pit composting, on-surface composting has clear advantages.

In addition to the challenges mentioned above, water harvesting will play an increasing role in smallholder agriculture under changing climate/weather conditions (Gebrechorkos et al. 2018). Hence, smallholder farmers need to find new strategies to adapt to climate change, not only with regard to soil nutrient recovery but also concerning water harvesting (FAO 2017b).

6   Conclusion

The use of farm waste has played a crucial role in traditional smallholder banana-coffee-based farming systems in Tanzania but lost attention since the demand for food and biofuels grew at a faster rate than agricultural production. However, the full potential of farm waste as a fertiliser and conditioner in sustainable soil management is not yet exhausted in traditional smallholder banana-coffee-based farming systems, as composting to our knowledge is not widely practised by all farmers despite efforts to revive it by farmer field schools and other initiatives. In order to keep pace with growing regional and global challenges, traditional composting practices must be adapted and integrated into soil and water conservation strategies (e.g., conservation tillage, terracing, water harvesting). Future research on on-surface composting should, therefore, focus on different compositions and amounts of macro- and micronutrients released from the different kinds of produced composts and should consider the nutrient demand of specific food crops.

We further conclude that future research should also focus on the questions: (1) how flows and stocks of nutrients within a farming system change with newly introduced composting practices; (2) what are the burdens smallholder
farmers, that do not successfully restore soil fertility through composting, have to

cope with; and (3) which further practices, for example as part of agroforestry approaches, need to be considered to transform degraded banana-coffee-based farming systems into long-term sustainable and intensive agricultural systems?

References

Akinnifesi, F. K., Sileshi, G., Ajayi, O. C., Chirwa, P. W., Kwesiga, F. R., & Harawa, R. (2008). Contributions of agroforestry research and development to livelihood of smallholder farmers in southern Africa: 2. Fruit, medicinal, fuelwood and fodder tree systems. *Agricultural Journal, 3*(1), 76–88.

Baijukya, F. P. (2004). *Adapting to change in banana-based farming systems of Northwest Tanzania: The potential role of herbaceous legumes.* Wageningen: Wageningen University.

Baijukya, F. P., & Steenhuijsen Piters, B. (1998). Nutrient balances and their consequences in the banana-based land use systems of Bukoba district, Northwest Tanzania. *Agriculture, Ecosystems & Environment, 71*(1–3), 147–158.

Baijukya, F. P., de Ridder, N., Masuki, K. F., & Giller, K. E. (2005). Dynamics of banana-based farming systems in Bukoba district, Tanzania: Changes in land use, cropping and cattle keeping. *Agriculture, Ecosystems & Environment, 106*(4), 395–406.

Baulme, R. (1993). *Crop water and irrigation requirements: Report to CES.* Salzgitter GmbH, Lingen: Consulting Engineers.

CARE and Overseas. (1994). *Refugee inflow into Ngara and Karagwe Districts, Kagera Region, Tanzania.* Environmental Impact Assessment, London, UK.

CGIAR-CSI. (2017). *SRTM 90m digital elevation data.* Accessed 11 Oct 2017. http://srtm.csi.cgiar.org/SELECTION/inputCord.asp

Copeland Reining, P. (1967). *The Haya. The Agrarian system of a Sedentary people.* PhD thesis, The Faculty of the Social Sciences, Department of Anthropology, The University of Chicago, Chicago, Illinois, USA.

Dancer, H. (2015). *Women, land and justice in Tanzania.* Eastern Africa series, James Currey.

FAO. (2006). *World reference base for soil resources: A framework for international classification, correlation, and communication* (Vol. 103: World soil resources reports), Rome: Food and Agriculture Organization of the United Nations (FAO).

FAO. (2015). *Agroecology for food security and nutrition.* In Proceedings of the FAO International symposium, Rome, Italy.

FAO. (2017a). *Agricultural transformation in Africa. The role of natural resources.* Nature & Faune (31st ed.). Accra, Ghana.

FAO. (2017b). *Sustainable land management (SLM) in practice in the Kagera Basin: Lessons learned for scaling up at landscape level.* Results of the Kagera Transboundary Agro-ecosystem Management Project (Kagera TAMP) 1(st ed.). Food and Agriculture Organization of the United Nations, Rome, Italy.

Garrity, D., Dixon, J., & Boffa, J.-M. (2012). *Understanding African farming systems. Science and Policy Implications.*

Gebrechorkos, S. H., Hülsmann, S., & Bernhofer, C. (2018). Changes in temperature and precipitation extremes in Ethiopia, Kenya, and Tanzania. *International Journal of Climatology, 4*(2), 1–13.

Gliessman, S. R. (2015). *Agroecology: The ecology of sustainable food systems.* Boca Raton: CRC Press.

Gliessman, S. R. (2016). Transforming food systems with agroecology. *Agroecology and Sustainable Food Systems, 40*(3), 187–189.

Hemp, C., & Hemp, A. (2008). *The Chagga Homegardens on Kilimanjaro. Diversity and refuge function for indigenous fauna and flora in anthropogenically influenced habitats in tropi-*

cal regions under global change on Kilimanjaro, Tanzania. In IHDP, Mountainous Regions: Laboratories for adaptation: Magazine of the International Human Dimensions Programme on Global Environmental Change, IHDP Update, Issue 2, IHDP, Bonn, Germany, pp. 12–17.

Katoke, I. K. (1970). *The making of the Karagwe Kingdom.* The historical Association of Tanzania, Dar es Salaam, Tanzania.

Kimaro, D. N., Kilasara, M., Noah, S. G., Donald, G., Kajuri, K., & Deckers, J. A. (1999). *Characteristics and management of soils located on specific landform units in the northern slopes of Uluguru Mountains, Tanzania, SUA.* In Proceedings of the Fourth annual conference of the faculty of agriculture on agricultural research challenges for the 21st century, Morogoro, Tanzania, Sokoine University of Agriculture (SUA), Morogoro, Tanzania.

Kimaro, D. N., Deckers, J. A., Poesen, J., Kilasara, M., & Msanya, B. M. (2005). Short and medium-term assessment of tillage erosion in the Uluguru Mountains, Tanzania. *Soil and Tillage Research, 81*(1), 97–108.

Kimaro, D. N., Terengia, S., Kihupi, N. I., Mtakwa, P. W., Poesen, J., & Deckers, J. (2011). *Conservation agriculture in the highlands of Tanzania under a Coffee-Banana-Agroforestry Farming System.* In Training Manual for Small Scale Farmers, Conservation Agriculture for a Restored Environment (NSSCP-CARE) Project, Morogoro, Tanzania.

Krause, A., Kaupenjohann, M., George, E., & Koeppel, J. (2015). Nutrient recycling from sanitation and energy systems to the agroecosystem- ecological research on case studies in Karagwe, Tanzania. *African Journal of Agricultural Research, 10*(43), 4039–4052.

Krause, A., Nehls, T., George, E., & Kaupenjohann, M. (2016). Organic wastes from bioenergy and ecological sanitation as a soil fertility improver: A field experiment in a tropical andosol. *The Soil, 2*(2), 147–162.

Lichtfield, J., & McGregor, T. (2008). *Poverty in Kagera, Tanzania: Characteristics, causes and constraints.* Brighton, UK.

Map Library. (2007). *Africa and Tanzania*, Shapefile.

Masawe, J. L. (1992). Farming systems and agricultural production among small farmers in the Uluguru Mountain area, Morogoro region, Tanzania. *African Study Monographs, 13*(3), 171–183.

Msita, H. B., Mtakwa, P. W., Kilasara, M., Kimaro, D. N., Msanya, B. M., Ndyetabula, D. K., Deckers, J. A., & Poesen, J. (2010). *Effect of conservational tillage on soil loss and plant nutrient status on vegetable yield, Northern Slopes of Uluguru Mountains, Morogoro, Tanzania.* Paper presented at Proceedings of the workshop on information sharing among soil and water management experts from SADC Universities.

Ndege, J., Steenhuijsen Piters, B., Nyanga, A., & Ngimbwa, L. (1995). *Diagnostic survey of Karagwe district.* Tanzanian-Netherlands Farming Systems Research Project, Lake Zone, Royal Tropical Institute (KIT), Amsterdam, Karagwe District Rural Development Programme, Ari Maruku, Bukoba, Ari Ukiriguru, Mwanza, Amsterdam, Tanzania and The Netherlands.

Reetsch, A., Feger, K.-H., Schwärzel, K., Dornack, C., & Kapp, G. (2020a). Organic farm waste management in degraded banana-coffee-based farming systems in north-West Tanzania. *Agricultural Systems.* (accepted).

Reetsch, A., Feger, K.-H., Schwärzel, K., & Kapp, G. (2020b). Transformation of degraded banana-coffee-based farming systems into multifunctional agroforestry systems – A mixed methods study from NW Tanzania. *Agricultural Systems.* (under review).

Rugalema, G. H., Okting'ati, A., & Johnsen, F. H. (1994). The homegarden agroforestry system of Bukoba district, North-Western Tanzania. 1. Farming system analysis. *Agroforestry Systems, 26*(1), 53–64.

Schwärzel, K., Zhang, L., Avellan, T., & Ardakanian, R. (2017). The water-soil-waste Nexus in Sub-Saharan Africa: Potentials for increasing the soil productivity. In R. Lal (Ed.), *Encyclopedia of Soil Science.* Boca Raton: CRC Press.

Shetto, R. M. (1999). Indigenous soil conservation tillage systems and risks of animal traction on land degradation in Eastern and Southern Africa. In P. G. Kaumbutho & T. E. Simalenga (Eds.),

Conservation tillage with animal traction. Animal Traction Network for Eastern and Southern Africa (ATNESA) supported by French Cooperation, Namibia, Harare, Zimbabwe, pp. 67–73.

Tittonell, P., & Giller, K. E. (2013). When yield gaps are poverty traps. The paradigm of ecological intensification in African smallholder agriculture. *Field Crop Research, 143,* 76–90.

Touber, L., & Kanani, J. R. (1996). *Landforms and soils of Karagwe District.* Karagwe District Council, Karagwe District Rural Development Programme, Karagwe, Kagera region, Tanzania.

Trærup, S. L. M., & Mertz, O. (2011). Rainfall variability and household coping strategies in northern Tanzania: A motivation for district-level strategies. *Regional Environmental Change, 11*(3), 471–481.

URT. (2012). *National sample census of agriculture 2007/2008.* Regional Report – Kagera Region. Volume Vh, Dar es Salaam, Tanzania.

URT. (2016). *Kagera Region: Basic demographic and socio-economic profile.* 2012 Population and Housing Census, Kagera Profile.

URT. (2018). *Statistic data Tanzania.* Statistics for Development, Tanzania National Bureau of Statistics.

Wasige, J. E., Groen, T. A., Smaling, E., & Jetten, V. (2013). Monitoring basin-scale land cover changes in Kagera Basin of Lake Victoria using ancillary data and remote sensing. *International Journal of Applied Earth Observation and Geoinformation, 21,* 32–42.

Wekesa, A., & Jönsson, M. (2014). *Sustainable agricultural land management.* Vi Agroforestry, Regional Office East Africa, Kenya.

Permissions

The contributors of this book come from diverse backgrounds, making this book a truly international effort. This book will bring forth new frontiers with its revolutionizing research information and detailed analysis of the nascent developments around the world.

We would like to thank all the contributing authors for lending their expertise to make the book truly unique. They have played a crucial role in the development of this book. Without their invaluable contributions this book wouldn't have been possible. They have made vital efforts to compile up to date information on the varied aspects of this subject to make this book a valuable addition to the collection of many professionals and students.

This book was conceptualized with the vision of imparting up-to-date information and advanced data in this field. To ensure the same, a matchless editorial board was set up. Every individual on the board went through rigorous rounds of assessment to prove their worth. After which they invested a large part of their time researching and compiling the most relevant data for our readers.

The editorial board has been involved in producing this book since its inception. They have spent rigorous hours researching and exploring the diverse topics which have resulted in the successful publishing of this book. They have passed on their knowledge of decades through this book. To expedite this challenging task, the publisher supported the team at every step. A small team of assistant editors was also appointed to further simplify the editing procedure and attain best results for the readers.

Apart from the editorial board, the designing team has also invested a significant amount of their time in understanding the subject and creating the most relevant covers. They scrutinized every image to scout for the most suitable representation of the subject and create an appropriate cover for the book.

The publishing team has been an ardent support to the editorial, designing and production team. Their endless efforts to recruit the best for this project, has resulted in the accomplishment of this book. They are a veteran in the field of academics and their pool of knowledge is as vast as their experience in printing. Their expertise and guidance has proved useful at every step. Their uncompromising quality standards have made this book an exceptional effort. Their encouragement from time to time has been an inspiration for everyone.

The publisher and the editorial board hope that this book will prove to be a valuable piece of knowledge for researchers, students, practitioners and scholars across the globe.

List of Contributors

Federico Varalta and Jaana Sorvari
Aalto University, Espoo, Finland

Ashootosh Mandpe, Sweta Kumari and Sunil Kumar
Council of Scientific & Industrial Research (CSIR)-National Environmental Engineering and Research Institute (CSIR-NEERI), Nagpur, India

Luis Fernando Marmolejo-Rebellón and Patricia Torres-Lozada
Faculty of Engineering, Universidad del Valle, Cali, Colombia

Edgar Ricardo Oviedo-Ocaña
School of Civil Engineering, Universidad Industrial de Santander, Bucaramanga, Colombia

Hiroshan Hettiarachchi, Serena Caucci, Cristian Rivera Machado and Lulu Zhang
United Nations University (UNU-FLORES), Dresden, Sachsen, Germany

Johan Bouma
Wageningen University, Wageningen, The Netherlands

Warshi S. Dandeniya
Department of Soil Science, Faculty of Agriculture, University of Peradeniya, Peradeniya, Sri Lanka

Laura Giagnoni
Department of Agriculture, Food, Environment and Forestry (DAGRI), University of Florence, Florence, Italy

Tania Martellini, Roberto Scodellini and Alessandra Cincinelli
Department of Chemistry, University of Florence, Sesto Fiorentino, Italy

Giancarlo Renella
Department of Agronomy, Food, Natural Resources, Animals and Environment, University of Padua, Legnaro, Italy

Dzidzo Yirenya-Tawiah, Ted Annang, Benjamin Dankyira Ofori, Benedicta Yayra Fosu-Mensah, Elaine Tweneboah-Lawson, Cecilia Datsa and Christopher Gordon
Institute for Environment and Sanitation Studies, University of Ghana, Accra, Ghana

Richard Yeboah and Kwaku Owusu-Afriyie
Management for Development Foundation Training and Consultancy, Accra, Ghana

Benjamin Abudey
Ga-West Municipal Assembly, Local Government Office, Accra, Ghana

Ted Annan
Ministry of Food and Agriculture, Accra, Ghana

Ariane Krause
Leibniz Institute of Vegetable and Ornamental Crops (IGZ), Großbeeren, Germany

Anika Reetsch
Institute for Integrated Management of Material Fluxes and of Resources (UNU-FLORES)
Sebastian Kolowa Memorial University, Lushoto, Tanzania

Didas Kimaro
Sebastian Kolowa Memorial University, Lushoto, Tanzania

Karl-Heinz Feger
Technische Universität Dresden, Dresden, Germany

Kai Schwärzel
Thünen Institute of Forest Systems, Eberswalde, Germany

Index

A

Additives, 128-130, 132-133, 138, 140, 148-149

Aeration, 28-29, 86, 133, 136, 143, 150

Agricultural Sector, 61-62, 65-66, 71, 75, 86, 91-92, 145

Agricultural Waste, 22, 85, 97, 103-105, 108, 124

Anaerobic Digestion, 4, 17, 22, 35, 76, 95, 110, 136, 176, 185, 205

Antibiotic Resistance, 17, 62, 72, 98, 108, 112-115, 120, 122-124

B

Biochar, 1, 4-10, 12, 14-20, 98, 106, 122, 129, 131-133, 148-152, 176-178, 183, 185, 187, 192-193, 196-200, 202-207

Biodegradable Waste, 21-22, 27-30, 32-35, 102, 106, 109-110, 121, 136, 151

Biogas, 4, 11-12, 17, 35, 76-78, 136, 176-178, 181-183, 185-196, 201, 205-207

Biogas Slurry, 176-178, 182, 185-196, 201

Bulking Agent, 29, 132-133, 138, 150-151

C

Circular Economy, 1-2, 14, 17, 61, 68, 154

Climate Change, 25, 40, 57, 63, 67, 137, 154, 177, 192, 201-202, 209, 218, 220, 224

Compost, 1, 7-10, 12, 14-19, 24-26, 28, 30-40, 42-44, 49-51, 57, 59-76, 82, 85-87, 91-93, 97-108, 215, 217, 222, 224

Compost Application, 19, 61, 99-100, 103, 122, 169

Compost Production, 65, 99, 101-102, 105-108, 111, 114, 121, 149, 154-155

Composting Plant, 33, 36, 75, 89

Composting Practices, 29, 60, 208-211, 215-217, 220, 222-224

Composting Process, 8-10, 14, 21, 24-25, 28, 30, 37, 43, 50-52, 54-55, 57, 59-61, 69, 83-84, 94, 97, 99, 199-200, 215

Composting Technique, 27, 29, 31-32, 39, 111

Composting Technology, 12, 31, 38-40, 73, 101, 127

Crop Residue, 68, 73

D

Dredged Sediment, 151

E

Ecosystem, 2, 6, 14, 62-63, 69, 71-72, 119, 201, 225

F

Farm Waste, 97, 101, 107, 167, 208-210, 215-218, 224, 226

Feedstock, 5, 7-8, 10, 12, 14-15, 19, 26, 35, 126-127

Final Disposal, 44, 49, 57, 64, 75-79, 81, 96

Food Chain, 16, 18, 69, 97-98, 115-116, 118, 120

Food Waste, 3, 29-30, 47, 51, 58-59, 94, 128, 131, 138, 149, 152, 167, 215

G

Green Waste, 3, 33, 40, 127-128, 131-132, 134, 140-141, 143, 145, 147, 149, 151-152

Greenhouse Gas, 1, 16-17, 24, 41, 63, 67, 75-76, 96, 122, 129, 132-133, 137, 140, 147, 150, 154, 194, 202, 206-207

Greenhouse Gas Emission, 16, 63, 129, 133, 137

Growing Media, 127, 140-141, 143, 146-147

H
Holistic Approach, 1-8, 12-16
Humification, 10, 20, 24-25, 131, 141, 199

I
Integrated Management, 62, 67-68, 75, 91, 94, 140, 147, 208

K
Kitchen Waste, 33, 41, 97, 106, 132, 152, 181, 197, 199, 224

L
Land Degradation, 62-63, 67, 145, 209, 221, 223, 226
Landfill, 24, 31-32, 35, 37, 44, 48-49, 57, 81, 92-93, 116, 123, 127, 136-137, 150, 155, 171
Landfilling, 26, 41, 64, 77-78, 86, 136-137, 140, 154
Leachate, 76, 91, 103, 106, 110-111, 114, 119

M
Municipal Solid Waste, 4, 21-22, 40-44, 58-59, 61-62, 72, 74-76, 95-96, 98, 108, 122-124, 128, 130, 132, 175, 202

N
Nutrient Content, 26, 53, 55, 124, 129, 197
Nutrient Management, 19, 97-100, 102, 115, 125, 220
Nutrient Recycling, 14, 60-61, 64-65, 67, 76, 192, 196, 204, 209, 226

O
Organic Carbon, 37, 51, 86-87, 99, 104, 131, 141, 187, 193, 207
Organic Fertilizer, 4

Organic Material, 75, 196, 208-210, 216, 218
Organic Matter, 2-3, 5-6, 8-10, 14-16, 19, 24, 35, 41, 44, 51, 58, 61, 63, 69-71, 73, 99, 131-132, 134, 138, 143, 176-177,224
Organic Pollutant, 119
Organic Waste, 1-7, 9, 11-17, 25, 28, 30, 32, 36-39, 41, 43-44, 46-54, 57, 60-61, 64-67, 69-73, 75-78, 81-86, 88-93, 216
Organic Waste Composting, 43, 51-53, 57, 60-61, 67, 69, 75-76, 81, 83, 89

P
Plant Nutrient, 102, 116, 125, 195, 221, 226
Poultry Litter, 30, 111-114, 122-123, 125, 151, 202
Pyrolysis, 4-5, 7-8, 12, 19, 22, 74, 148, 204

R
Recycling, 1-6, 11, 13-16, 25, 27, 41-42, 44, 46, 48-49, 58-61, 64-67, 70, 72-73, 76-79, 82, 88, 92, 95-96, 98, 101, 226
Reducing, 1-2, 5, 8-10, 16, 18, 25, 27, 34, 37, 63, 91, 98, 100, 102, 105, 114, 123, 132, 151, 185, 197, 200, 213

S
Sanitation, 5, 43, 46-47, 56, 67, 79, 127, 129, 131, 133-134, 153, 155, 157, 172, 174, 176-179, 181, 185-186, 226
Sewage Sludge, 8, 10, 12, 16, 18, 20, 25-26, 41, 61, 69, 74, 124, 127-129, 132-134, 136, 149, 151
Soil Amendment, 4, 7, 116, 126, 128, 140, 146-147, 192, 202
Soil Erosion, 19, 63, 72-73, 180, 186, 208-209, 214, 218, 220-221
Soil Fertility, 19, 21-22, 32, 37, 44, 72, 98-99, 101, 116, 124, 149, 157, 163, 176-178, 180-181, 185-187, 226
Soil Fertility Management, 99, 176-178, 180-181, 186-187, 191-192, 197,

Soil Management, 60, 69, 71, 123, 157, 180, 185, 188, 190, 192, 194, 224

Soil Organic Matter, 6, 14, 44, 61, 63, 71, 176-177, 207, 218, 222

Soil Quality, 7, 16, 63, 70-71, 150, 163, 172, 186, 202

Solid Waste, 4, 18, 21-22, 25, 30-31, 36-37, 40-44, 48, 57-59, 61-62, 64, 72-76, 79, 84, 90, 95-96,174-175, 202

Solid Waste Management, 22, 25, 30-31, 36-37, 40-42, 48, 57-58, 72-75, 79, 84, 95-96, 102-103,160, 174-175

Source Separation, 26, 44, 47, 49, 65, 75-76, 82-84, 88, 90-91, 94

Substrate, 4, 7, 11-12, 22, 24, 28-30, 48, 50-51, 53, 55, 57, 85, 194

Sustainability, 1-4, 6, 11-18, 21, 38, 50, 64, 70, 91, 95, 122, 124-125, 127, 146-148, 165, 173, 206

Sustainable Development, 1, 3, 15, 18-19, 40, 60-61, 67, 71-73, 89, 149, 154, 177, 205

T

Terra Preta, 1, 6-8, 10, 17, 177, 190, 196, 203

Thermophilic Phase, 8, 24, 127, 131, 134, 138, 141, 145, 206

V

Vermicomposting, 18, 27-29, 31, 36, 49-50, 58, 152

W

Waste Collection, 34, 36, 48, 57, 60, 66, 70, 78, 82-83, 90, 93, 105, 109, 111, 154, 158-159, 161, 166, 169, 171

Waste Generation, 22, 33, 39, 50, 60, 68, 70, 80, 93, 102, 109, 154-155

Waste Management, 1, 3-6, 18-19, 21-22, 25, 27, 30-33, 35-42, 44, 46-48, 50, 57-59, 205, 226

Waste Recovery, 49, 73, 78-79, 82, 85, 87-88, 90, 154-155

Waste Segregation, 34, 39, 154, 156, 160-161, 167, 173-174

Waste Stream, 21, 26, 49, 107, 154-155, 167

www.ingramcontent.com/pod-product-compliance
Lightning Source LLC
Chambersburg PA
CBHW050124240326
41458CB00122B/1227